U0343161

超越 STUDIO
SUPER 设计课

向畅颖　金梦潇　郑权一 / 编著

作品集炼成记

The Art of Portfolio-Key to World-Leading Universities

建筑、城规、景观 名校留学必修书

本书是关于留学作品集的指导书。对建筑学、城市规划、景观、室内设计等设计热门专业留学作品集的构成、准备要点、优劣的评定原则、不同学校的要求及优秀作品集的水平都有着详细介绍。全书包括什么是留学作品集、留学作品集的准备、留学作品集要素详解、优秀作品集案例及院校要求详解四部分内容，并配有录取名校学生的相应作品集配图。本书是有留学需求的设计专业学生和设计师的绝佳选择。

图书在版编目（CIP）数据

作品集炼成记：建筑、城规、景观名校留学必修书/向畅颖，金梦潇，郑权一编著. —北京：机械工业出版社，2018.3

（超越设计课）

ISBN 978-7-111-59393-5

Ⅰ.①作⋯ Ⅱ.①向⋯②金⋯③郑⋯ Ⅲ.①建筑设计—作品集—中国—现代 Ⅳ.①TU206

中国版本图书馆CIP数据核字（2018）第047854号

机械工业出版社（北京市百万庄大街22号 邮政编码100037）
策划编辑：赵 荣 责任编辑：赵 荣 张维欣
责任校对：黄兴伟 封面设计：鞠 杨
责任印制：常天培
北京联兴盛业印刷股份有限公司印刷
2018年7月第1版第1次印刷
184mm×260mm·17.5印张·2插页·438千字
标准书号：ISBN 978-7-111-59393-5
定价：119.00元

凡购本书，如有缺页、倒页、脱页，由本社发行部调换

电话服务 网络服务
服务咨询热线：010-88361066 机工官网：www.cmpbook.com
读者购书热线：010-68326294 机工官博：weibo.com/cmp1952
　　　　　　　010-88379203 金 书 网：www.golden-book.com
封面无防伪标均为盗版 教育服务网：www.cmpedu.com

序
PREFACE

  本书出自为建筑梦执着的向畅颖、郑权一和我。

  尽管我们仨是同窗，但第一次熟悉向畅颖先生，却是一个充满阳光的暑假午后。那时我和郑权一先生刚从台湾作为建筑交换生提前返回母校。那是一个英气的剪影，向先生独自一人坐在哈尔滨工业大学建筑学院专教的最中央，面前摞着比他还高的复习参考书，时而奋笔疾书，时而大声朗读准备着即将考的德福、托福、GRE、雅思和建筑学留学必备的作品集。后来的故事大家应该都会猜到，向先生这位大神级的人物，申请了24所全世界建筑顶级名校并全部收到offer，而充满了浪漫色彩的他，去了童话王国丹麦。

  我们开始一起创业，是我和向先生一起在哈尔滨工业大学－加州大学伯克利分校可持续都市发展联合研究中心做研究助理，那时候郑博士已经建立巅峰建筑学社，辅导了很多考研的学生，出了两本书《巅峰建筑快题设计实例教程150》《水晶石技法——建筑设计手绘实例教程》。由于接触的学生越来越多，咨询留学的学生也越来越多，经常有学弟学妹跑到我们办公室和向博士咨询留学的相关事宜。那时候全国还没有专门辅导设计类作品集留学的机构，我们仨作为第一个吃螃蟹的人开始为全国广大设计类留学的同学辅导及咨询。

  一年前的冬天，向先生归国探亲，我们又回到母校，坐在温暖的咖啡厅里，准备筹划分享给全国有建筑留学梦想的同学都能学习的书。这本书里承载着很多学生的梦与汗水，承载着很多成功的经验与分享，承载着很多的年华与诚意，我们将申请世界名校的优秀作品集案例毫无保留地展示出来，希望给更多莘莘学子更全面、更详细、更直观的学习机会。同时，也希望那些还在为留学迷茫和选择学校犹豫的同学，通过我们精心的信息收集，指点迷津，确定奋斗的方向。

  每个准备留学的学生都对自己的未来有着一个美好的规划，你们不甘于平庸，愿意通过自己的努力证明自己，无论是希望去更广阔的平台施展自己的才能，还是希望能学成归国报效祖国、感恩父母；无论是希望一边学习一边游学增长阅历见识，还是希望能交到更优秀的挚友和寻觅到美好的爱情，希望您都能以梦为马，不忘初心，勇敢前行。作为一名建筑师，最幸福的事就是从此赋予了自己一双建筑师的眼睛来观察这个世界，就像一种信仰，像信仰宗教一样对建筑多了一份虔诚，那么最后会发现，世间唯有建筑和梦想不可辜负。

  在这里感谢培养出我们仨的令人敬仰和爱戴的博士生导师们：南方科技大学的郑春苗老师、哈尔滨工业大学的张姗姗老师、赵晓龙老师、李桂文老师及邹广天老师。师恩难忘。

  最后，祝福我们的学生们及正在阅读本书的您：

    前程似锦，阅尽千帆，走出半生，归来依旧是少年。

金梦潇
于深圳

PREFACE 前言

  这是一本关于留学作品集的指导书，是一本深入浅出，从经验中来到实践中去的留学作品集参考书。本书的源起来自笔者、金梦潇女士和郑权一先生在哈尔滨冰天雪地温暖的咖啡屋里某个畅谈的下午。那时我们专注设计类专业的留学辅导已经有数年，我们辅导的学生获得众多全球顶级名校的 offer 让我们深感欣慰。但我们在教学过程中发现在作品集入门阶段大多数学生都有很多的困惑和共性的问题，而市面上关于留学作品集的专业指导书凤毛麟角，具有时效性和全面性同时又具体到每一个细节的书更是几乎没有。因此我们感到需要给学生们写一本扫清迷雾的留学作品集指导书，给每一个即将勇敢踏上这翻山越岭旅程的学子一本指南手册，把我们宝贵的一手经验分享给大家，助大家飞得更高、更远，更顺利地实现自己的梦想。

  从笔者开始构思到下笔成书历时一年有余，整体一气呵成，一些细节数易其稿，每当有最新的重要留学信息我们都会添加进相应的章节。这样也保证了本书的时效性。本书面向的读者范围很广，从设计专业低年级对留学有兴趣的同学到已经工作想要再深造的设计界同行们都可以借鉴。阅读本书，大家可以对设计留学作品集的构成、准备要点、优劣的评定原则、不同学校的要求以及优秀作品集的水平都有详细的了解。本书涵盖建筑、景观建筑学、城市设计、室内设计等主要留学设计类热门专业的作品集介绍，同时将作品以庖丁解牛的方式详细分解，每一部分、每一类图纸都结合我们的经验讲解，并配有录取名校学生的相应作品集配图。这样可以让对作品集一头雾水不知如何下手的同学做到心中有数，有的放矢地设定目标。也可以让设计水平较好的同学们更有针对性地提高自己的设计作品，精确瞄准心仪的名校，节省精力和时间，提高录取率。

  本书是一本抛砖引玉的书，知识无界，分享能增强知识力量本身。我们深切地希望这本书可以帮助广大深深热爱设计、执着追求自己梦想的同学们。让同学们在制作作品集的旅途上不断变得强大，练成属于自己的绝世武功，得到理想名校的录取通知书。

向畅颖
于丹麦奥尔堡

目录

序
前言

第一章　什么是留学作品集 ·· 01
第一节　解读留学作品集 ·· 02
　　一、什么是留学作品集 ·· 02
　　二、留学作品集与其他作品集的区别 ·· 02
　　三、留学作品集适用人群 ··· 05
第二节　留学作品集在申请中的分量 ·· 11
　　一、设计类留学申请的特殊性 ·· 11
　　二、专业申请者的困惑：孰轻孰重？语言，文书，作品集？ ··························· 12
　　三、转专业申请者的终极兵器：打开名校大门的金钥匙 ································· 14
第三节　什么是好的留学作品集 ·· 16
　　一、拨开重重迷雾，申请者的误区解读 ·· 16
　　二、优秀作品集必备的要素 ··· 17

第二章　留学作品集的准备 ·· 23
第一节　留学作品集的准备周期和时间安排 ··· 24
　　一、准备周期 ··· 24
　　二、准备时间安排 ·· 24
　　三、作品集和语言考试的时间分配 ··· 25
第二节　留学作品集准备的技能及水平 ·· 27
　　一、软件水平要求 ·· 27
　　二、模型制作 ··· 40
　　三、手绘表达 ··· 42
第三节　留学作品集的内容准备 ·· 47
　　一、什么样的内容适合留学作品集 ··· 47
　　二、作品集任务书及内容筛选 ·· 47
　　三、结合院校以及自身背景制定作品集框架 ·· 49

第三章 留学作品集要素详解 ·············51

第一节 作品集方案要素 ·············52
一、平面图 ·············52
二、立面图 ·············54
三、剖面图 ·············55
四、总图 ·············57
五、分析图 ·············59
六、效果图 ·············62
七、其他 ·············64

第二节 作品集的表达 ·············66
一、单个作品表达 ·············66
二、全套作品集表达 ·············66

第三节 作品集排版及配色 ·············67
一、作品集排版 ·············67
二、作品集配色 ·············68

第四章 优秀作品集案例及院校要求详解 ·············71

第一节 优秀作品集案例展示 ·············72
一、建筑类专业方案 ·············72

● 别墅小住宅类设计作品 ·············72
  北京白塔寺院落改造设计（米兰理工大学）·············72
  香港摄影师工作室兼住宅设计（香港中文大学）·············80
  可持续漂浮住宅设计（新南威尔士大学）·············85
  模块化住宅设计（墨尔本大学）·············88

● 茶室花房类设计作品 ·············92
  花房设计（米兰理工大学）·············92
  茶室设计（谢菲尔德大学）·············98

● 幼儿园及小学类建筑设计作品 ·············104
  香港 L——shape 幼儿园设计（香港中文大学）·············104
  小学设计（谢菲尔德大学）·············108

● 博物馆展览馆类建筑设计 ·············116
  窑洞博物馆设计（瑞典皇家理工学院）·············116
  厦门大学科学博物馆设计（米兰理工大学）·············120
  记忆博物馆设计（谢菲尔德大学）·············128

前海定制博物馆设计（瑞典皇家理工学院）……………………………………136
- 校园建筑设计……………………………………………………………………141
  校园共享广场设计（伦敦大学学院，简称UCL）………………………………141
  校园港口——燕山大学学生活动中心设计（米兰理工大学）……………………147
  燕山大学建筑系馆设计（米兰理工大学）………………………………………153
  健身馆建筑设计（英国爱丁堡大学）……………………………………………158
- 旧建筑改造类设计………………………………………………………………162
  台湾猫咪之家——工厂改造设计（瑞典皇家理工学院）…………………………162
  造纸工厂再生——城市天桥展览建筑设计（香港中文大学）……………………166
  哈尔滨铁路局城市设计（米兰理工大学）………………………………………172
- 高层类建筑设计…………………………………………………………………177
  旋转空中花园设计（伦敦大学学院）……………………………………………177
  商住综合高层设计（墨尔本大学）………………………………………………181
  高层医院设计（墨尔本大学）……………………………………………………184
  高层胶囊公寓设计（曼彻斯特大学）……………………………………………188
  秦皇岛东北大学校园高层设计（米兰理工大学）………………………………193

二、城市规划类专业方案……………………………………………………199

- 滨水空间城市设计………………………………………………………………199
  松花江滨水空间可持续再生设计（瑞典皇家理工学院）…………………………199
- 历史街区城市再生设计…………………………………………………………205
  厦门集美大社文化再生城市设计（米兰理工大学）……………………………205
  南京火车站改造城市设计（伦敦大学学院）……………………………………213

三、景观类专业方案………………………………………………………………219

- 校园景观设计……………………………………………………………………219
  科罗拉多大学校园外道路景观设计（米兰理工大学）……………………………219
  校园景观步行系统设计（米兰理工大学）………………………………………224
- 公园景观设计……………………………………………………………………228
  公园景观修复设计（米兰理工大学）……………………………………………228
- 城市宗地景观改造………………………………………………………………233
  沈阳工业区景观再设计（米兰理工大学）………………………………………233
- 休闲景观设计……………………………………………………………………239
  辽宁本溪温泉景区景观设计（米兰理工大学）…………………………………239
  米兰理工可持续建筑及景观硕士　王雨歆………………………………………243

四、室内设计类专业方案…………………………………………………………246

- ● 住宅室内设计 ·· 246
  - 公寓室内设计（林肯大学）··· 246
- ● 餐饮空间室内设计 ·· 248
  - 哈尔滨中东铁路住宅改造——咖啡厅设计（威斯特斯特大学）······· 248
- ● 幼儿园室内设计 ··· 251
  - 幼儿园室内设计（伯明翰城市大学）··· 251

第二节　不同国家院校作品集的要求要点（境外一流名校）······················ 252

一、英国 ··· 252
- ● 伦敦大学学院作品集要求 ·· 252
- ● 谢菲尔德大学作品集要求 ·· 253
- ● 爱丁堡大学作品集要求 ··· 254
- ● 曼彻斯特大学作品集要求 ·· 255

二、澳大利亚 ·· 256
- ● 悉尼大学作品集要求 ·· 256
- ● 新南威尔士大学作品集要求 ··· 257
- ● 墨尔本大学作品集要求 ··· 257

三、美国 ·· 258
- ● 麻省理工学院作品集要求 ·· 258
- ● 哈佛大学作品集要求 ·· 259
- ● 宾夕法尼亚大学作品集要求 ··· 260

四、北欧 ·· 261
- ● 瑞典皇家理工学院作品集要求 ·· 261
- ● 阿尔托大学作品集要求 ··· 262
- ● 意大利米兰理工大学作品集要求 ··· 262
- ● 荷兰代尔夫特理工大学作品集要求 ·· 264

五、中国 ·· 265
- ● 香港大学作品集要求 ·· 265

后　记 ·· 266

CONTENTS

Chapter 1　What is portfolio for study abroad

1. Decipher the portfolio

2. Important roles of portfolio in application

3. The characters of a good application portfolio

Chapter 2　Preparation of portfolio

1. Preparation period and time schedule

2. Required skills and level for portfolio

3. Preparation for the portfolio content

Chapter 3　Detailed introduction of portfolio elements

1. Design concepts elements

2. Expression of portfolio

3. Portfolio layout and color selection

Chapter 4　Outstanding portfolio exemplars and requirements by different universities

1. Outstanding architectural portfolio exemplars

(1) Architecture

- Villa/House category

    Reinventing the courtyard(Politecnico di Milano)

    Photography studio(The Chinese University of Hong Kong)

    Amphibious house(UNSW)

    Module container house(University of Melbourne)

- Tea house/Green house category

    Small community center and Green house(Politecnico di Milano)

    The tea room(University of Sheffield)

- Kindergarten/Primary School category

    Kindergarten—— 'L 'shape(The Chinese University of Hong Kong)

    Primary School(University of Sheffield)

- Museum/Gallery category

    Cave-dwelling museum (KTH)

    Science Museum Design(Politecnico di Milano)

    The Museum of Memory(University of Sheffield)

    Qianhai Subscription(KTH)

- Campus building category

    Communication playground(UCL)

    Campus harbor—Yanshan University student activity center(Politecnico di Milano)

    Pavilion Rebirth(Politecnico di Milano)

    Fitness room architecfured design(university of Edinburgh)

- Architecture Renovation category

    Cat's theater—Factory renovation in Taiwan(KTH)

    Bridge Complex (The Chinese University of Hong Kong)

    Urban Design of Harbin Railway Administration (Politecnico di Milano)

- High-Rise Building category

    Spiral Skyscraper(UCL)

    Mix-use Building(University of Melbourne)

    Hospital Design(University of Melbourne)

    Capsule Residential Tower(University of Manchester)

    New Campus of Northeast University (Politecnico di Milano)

(2) Urban Design

- Waterfront Urban Design

    Sustainable Urban Waterfront Redesign of Harbin(KTH)

- Historical Urban Area Revitalization

    JiMeiDaShe Culture Creative Center(Politecnico di Milano)

    Urban Design Renovation Project of Pukou Railway Station(UCL)

(3) Landscape Architecture

- Campus Landscape

　　　　Off the Rails(Politecnico di Milano)

　　　　Sustainable Campus Pedestrian Zone Design(Politecnico di Milano)

- Park Design

　　　　Tanan Park Renovation(Politecnico di Milano)

- Urban Brown Field Landscape Design

　　　　Industrial Abandoned Landscape Design(Politecnico di Milano)

- Leisure Landscape Design

　　　　Liaoning Tanggou Hot Spring Resort Design(Politecnico di Milano)

　　　　Village Landscape Design(Politecnico di Milano)

(4) Interior Design

- Residential Interior Design

　　　　Apartment Interior Design (University of Lincoln)

- Coffee Shop and Book Store Design

　　　　Sino-Russia Railway House Renovation (University of Westminster)

- Kindergarten Interior Design

　　　　Kindergarten Interior Design(Birmingham City University)

2. Requirement by different Universities(Elite Universities)

(1) UK

- UCL Requirement

　　　　University of Sheffield Requirement

　　　　University of Edinburgh Requirement

　　　　University of Manchester Requirement

(2) Australia

　　　　University of Sydney Requirement

　　　　UNSW Requirement

　　　　University of Melbourne Requirement

(3) U.S.A.

　　　　MIT Requirement

Harvard University Requirement

University of Pennsylvanian Requirement

(4) Nordic Countries

KTH Requirement

Alto University Requirement

Politecnico di Milano Requirement

TU Delft Requirement

(5) China

University of Hong Kong Requirement

第一章

什么是留学作品集
What Is Portfolio For Study Abroad

Decipher The Portfolio
解读留学作品集

第一节

一、什么是留学作品集

留学作品集（portfolio/work samples for studying abroad purpose），简而言之，就是设计类专业学生和从业人员用于留学深造或学术交流申请用的作品集。基本上所有的设计类专业都需要作品集，包括建筑学、城市规划、景观建筑学、室内设计、服装设计、家具设计、产品设计等。作品集是展示申请者设计水平和知识结构最直观的材料，是留学申请材料中最重要的组成部分，也是境外院校机构评审申请最看重的环节之一。因此，一份好的留学作品集能在申请中奠定举足轻重甚至弥补其他申请材料短板的重要作用。

二、留学作品集与其他作品集的区别

说到作品集，所有设计类专业的学生和从业人员都不陌生，从大一开始很多同学就已经开始从学长学姐或者老师口中听到这个名词。但是对作品集有一个准确完善的认识需要一个过程。作品集并不是简单地把过往的设计作业和参与设计项目组合在一起，也不是所有图纸的集合。而是一个非常有目标性、针对目标阅读人群的设计作品展示材料。去粗取精，去繁

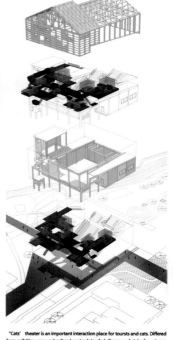

LAYER-ANALYSIS OF CATS' THEATER

"Cats' theater is an important interaction place for toursts and cats. Differed from exibition area and cultural center, lots of platforms and stairs for cats are design in cats' theater. In cats' theater, there are pleasant atmosphere for cat-human interatction. Tourists and cats are both audiance and actors.

PESPECTIVE SECTION PLAN OF CATS' THEATER

瑞典皇家理工学院建筑学专业录取的吴月设计作品（一）

求简，生动有趣而又细致是好的作品集的必备品质。针对阅读人群的不同，作品集又有很多细分，而且区别会很大，主要的归类有留学作品集、求职作品集、考研作品集等。每一类都有鲜明的特点和独特的要求，所以在制作的时候决不能一概而论，一份好的考研或者求职作品集如果用于留学申请，很可能会成为一个非常失败的反面例子。因此大家明确自己准备的作品集分类和所需注重的要点。在下面的章节里将会具体阐述留学作品集与其他作品集的主要区别。

（一）与研究生入学作品集的区别

对于准备读研的同学来说，研究生入学作品集一般都是不陌生的，无论是考研还是保研作品集都是非常重要的考核材料，特别是在第一次联系心仪院校的导师为自己建立良好的印象时，以及复试环节展示自己的能力时都有非常重要的作用。虽然同属针对 postgraduate（研究生）阶段学习的作品集，留学作品集和境内研究生面试的作品集却有很大不同，两者决不能一概而论或者通用。首先在留学作品集内容的选择上需要考虑到申请国家（地区）与院校的特色。相比而言，留学作品集更注重学术特色，符合国际上建筑发展的前沿变化，注重思维的思辨性和逻辑的严谨性。境内研究生入学作品集则一般符合我国大的建筑环境背景，可以将学术作品和实习工作等设计作品组合搭配。其次，在作品集的风格上境内外也有非常大的差别，比如说将境内考研作品集用于留学申请很可能会直接不通过，一本好的留学作品集用于境内考研如果调整不当也可能会不符合我国导师的预期。风格的迥异是由于阅读审核人员的不同而直接决定的。留学作品集的风格需要和主要申请目标国家（地区）、目标院校的风格相符，同时还需要满足具体的专业特色。例如申请瑞典皇家理工学院建筑学专业的作品集和申请米兰理工大学可持续建筑与景观专业的作品集就会有明显的风格区别。

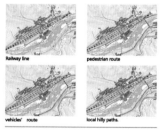

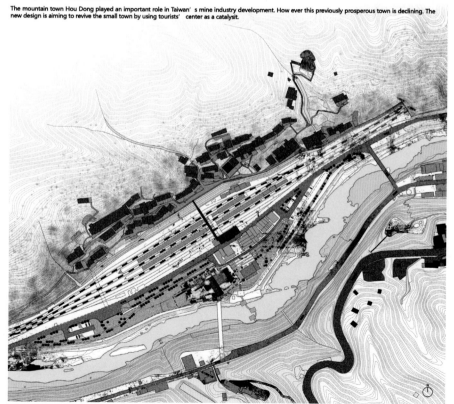

瑞典皇家理工学院建筑学专业录取的吴月设计作品（二）

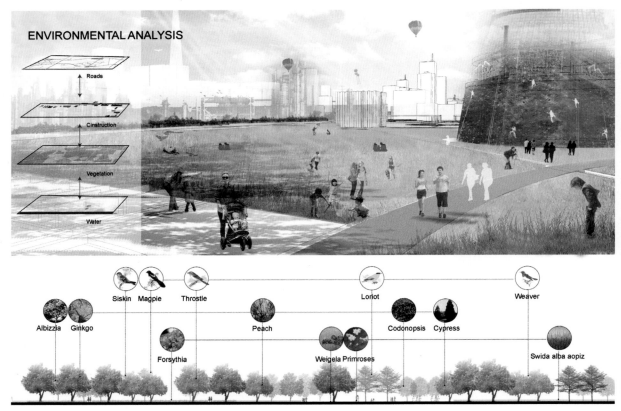

米兰理工大学可持续建筑及景观专业录取的王雨歆设计作品（一）

同理，用于考研或保研的作品集则也需要和报考院校风格接近，由于我国硕士实行导师制，作品集的风格也尽量和导师的风格特色相吻合。除了大家均能注意到的内容、风格、格式等区别以外，留学作品集和境内研究生入学作品集的另一个主要区别就是叙述性。一本好的留学作品集一定是能够独立成体系和引人入胜的，具有极强的故事叙述性的，而研究生入学作品集则往往不需要这些特征。换句话来说，相比研究生入学作品集，光有好设计并不一定就能构成优秀的留学作品集，故事要好，怎么讲述也很重要。具体的缘由其实很简单，申请者需要用留学作品集将自己的基本情况、设计水平、设计理念和特点以及知识结构通过图示和文字展示给评审阅读老师，这就要求有很出色的讲故事能力。而境内研究生入学作品集在复试或者面试过程中，考生可以在场解答和诠释自己的作品，并加以其他辅助材料佐证。这一点区别是很多申请境外院校的同学容易忽视的，正所谓不会讲故事的设计作品集不是好作品集，如何用图纸和文字讲出生动的故事，让评委老师眼前一亮，从万千作品集中挑出你的设计，兴致盎然手不释卷，则是需要着实下一番功夫的。

（二）与境内工作应聘作品集的区别

相对境内研究生入学作品集，留学作品集和境内申请工作应聘作品集另有一番区别。鉴于越来越多已工作同学也加入留学大军，以及许许多多在境外硕士毕业的同学希望毕业后顺利找到理想的工作，在这里我们介绍一下工作应聘作品集和留学作品集的主要区别，帮助大家避开不必要的误区。首先，工作应聘作品集相比留学作品集最大的区别就是两者所需表达的动机不同：工作应聘作品集是证明申请者是职位的完美候选人，从各个方面符合公司的预期，满足工作岗位的技能要求的证明材料；而留学作品集是吸引目标院校录取委员会兴趣，体现申请者个人兴趣风格，对设计理解和认知的展示材料。前者属于力求证明申请者符合公司预期（短期或长期）的能力证明材料，后者则是展示个人知

米兰理工大学可持续建筑及景观专业录取的王雨歆设计作品（二）

识结构、设计天赋和潜能的展示材料。前者是有的放矢的筛选过程，后者是不拘一格、伯乐慧眼识良驹的选拔过程。筛选与选拔，一字之差造就了工作应聘作品集和留学作品集完全不同的特征。具体来说，工作应聘作品集需要通过以往的学术，实际设计作品来体现申请者的设计能力，技术知识积累，团队协作能力，各类软件的使用和学习能力。HR（人力资源）会以挑剔的眼光看每个申请者的材料，短板是很致命的，他们是要寻找最合适的申请者，而不一定是最优秀的申请者。不同的岗位，不同的公司还会有完全不同的预期，比如，在方案投标型的公司职位招聘时，HR希望看到的是更具有创造力和竞赛投标经验的设计作品。而以实际项目为主的公司职位所预期的作品集会是更偏重于设计的合理性，能体现申请者实际工程经验，与各专业相协调的能力等。留学作品集作为应对选拔的展示材料则完全可以更学术化、个人风格化，这也是境外院校所希望看到的。目标院校录取委员会教授们更希望通过作品集了解到学生鲜明的性格特色，并不要求每个方面都面面俱到，完全满足学校的要求，短板的存在是允许的，更重要的是长处，足够好的长处完全可以弥补短板的缺陷。比如说有的同学对建筑结构或者技术知识掌握相对欠缺，和境外院校的要求有一定差距，但是如果设计理念和空间创造非常优秀，能得到评委老师们的欣赏，是完全有希望在第一轮挑选中就被教授们选中进入第二轮的。当然，我们建议留学作品集各方面都很全面，在尽量减少短板的基础上体现自身的长处，随着越来越多的同学志愿留学深造，留学申请的竞争也越来越激烈，境外院校对作品集的要求也不断水涨船高越来越严格，所以打算留学的同学们一定要早做准备。

三、留学作品集适用人群

谁最需要留学作品集呢？除了所有有留学意愿的同学们，所有打算在境外公司就职的毕业生和独立工作室的设计师们都有着相似的需求。

（一）留学的学生

留学的学生是留学作品集最大的适用人群，境外所有的优秀院校的设计专业都将作品集列为申请材料之一，无论形式是打印版、电子版还是网页链接，都少不了作品集这一项。建筑学、城市设计或城市规划、室内设计、景观建筑学、服装设计、产品设计等专业的学生留学都需要准备作品集。不同的国家，不同的学校对作品集都有具体的要求。申请者在准备之前需要仔细阅读所选学校所选专业对作品集的详细要求，包括必须或建议包含的内容、作品数量、作品集排版格式、页数和大小等。如果同时申请好几所院校，那么作品集需要视情况做一定的调整，进而满足不同院校的要求，这一点是申请者一定要注意的，如果事先未做详细的考虑和安排，用同一本作品集申请不同的院校很可能会掉进各种陷阱。例如，如果连格式和内容都和院校的明确要求不符，哪怕作品再好也会被评委老师视为不符合要求，不用心申请而拒绝。申请 UCL（伦敦大学学院）的作品集就需要打印出来邮寄，校方偏向 A4 横版的作品集，不建议打印在过厚的铜版纸上，这些都是学校有明确规定的。谢菲尔德大学、爱丁堡大学等英国名校也都对作品集格式有明确的要求。

墨尔本大学的作品集也明确规定需要 A3 横版作品集，总页数不超过 15 页。而英国很多院校的作品集需要页数在 30 页以上。

米兰理工大学、诺丁汉大学等名校也对作品集大小有着详细的规定。许多同学心仪的代尔夫特理工大学更是对作品集有着极其详细的规定，并且强调不符合要求的作品集学校有权不予受理。因此，准备留学的同学一定要首先了解好各个学校的要求，做足功课，切不可盲目地把一本作品集投递多个不同的院校，这样海投是非常忌讳的。大家一定要有针对性地用心制作作品集，比如米兰理工大学就青睐清新细腻的设计风格，申请米兰理工大学的同学就可以以清新风格为主。

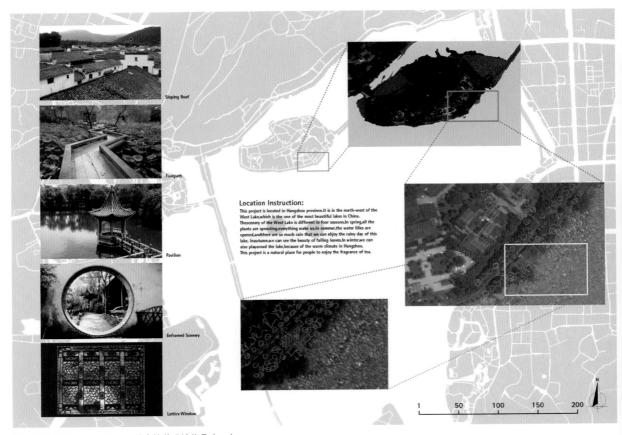

谢菲尔德大学建筑学专业录取的李怡萱设计作品（一）

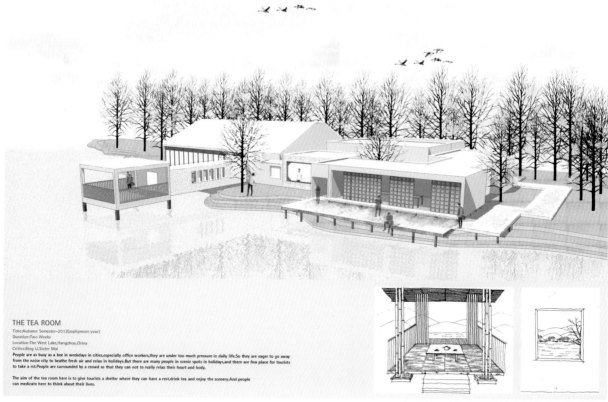

THE TEA ROOM

Time: Autumn Semester-2012 (sophomore year)
Duration: Two Weeks
Location: The West Lake, Hangzhou, China
Critics: Bing Li, Sister Mai

People are as busy as a bee in weekdays in cities, especially office workers, they are under too much pressure in daily life. So they are eager to go away from the noise city to beathe fresh air and relax in holidays. But there are many people in scenic spots in holidays, and there are few place for tourists to take a rst. People are surrounded by a crowd so that they can not to really relax their heart and body.

The aim of the tea room here is to give tourists a shelter where they can have a rest, drink tea and enjoy the scenery. And people can medicate here to think about their lives.

谢菲尔德大学建筑学专业录取的李怡萱设计作品（二）

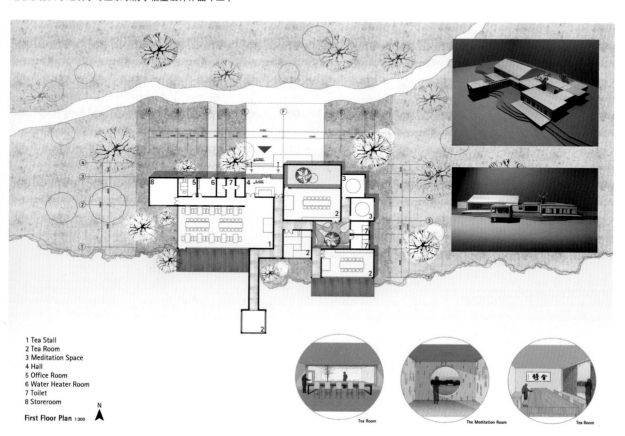

1 Tea Stall
2 Tea Room
3 Meditation Space
4 Hall
5 Office Room
6 Water Heater Room
7 Toilet
8 Storeroom

First Floor Plan 1:300

谢菲尔德大学建筑学专业录取的李怡萱设计作品（三）

Architecture Shopping Center

Design Concept

Hope to imitate the growing of plants through the form of building elevation, and bring feeling of closing nature to people in the boring modern life. At the same time, create a public site for the area, and change the social human environment in the area with the form of building magnetic field. For the building function, we strive to be the most effective and have the most economic benefits, and there is no conflict of stream lines. And provide a cultural center of daily activities and contacting for the residents in the area.

Rendering: In the aspect of traffic, hope to improve the crowded traffic in flyover, so the entrance is put in the east direction.

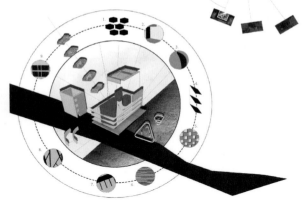

Rendering: At the same time, guide the sight of people to the halfspace, so as to be the areal center of public activities contacting.

A second strium allowed designer to re-imagine the built world as a natural community, quite literally. In The Meeting Park, the great expanse comes back down to the human scale, with stands of live bamboo and glass meeting rooms that evoke modern Chinese backyard conservatories.

墨尔本大学建筑学专业录取的刘宇恒设计作品（一）

Architecture Shopping Center

Design Concept

For the unique geographical location, the writer adopts flat and thick globe in the shape, to form a contract with the left tower body of Liaocheng TV Station; in the business arrangement and selling strategies, the writer also takes the shock of the modern gradual rising of e-commerce consumption on the traditional retail business except the supermarket function that itself has.

Rendering: arranges the Characteristic experience stores with strong theme and sense of belonging it does not exist in the surrounding commercial synthesizes.

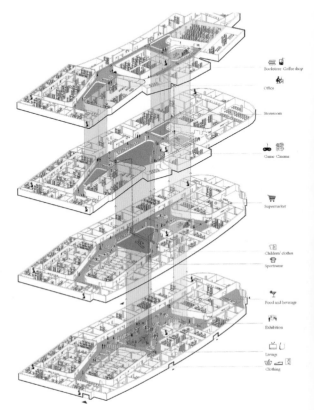

Rendering: Building is not only the building itself that designed on the red lines of lands, and the horizon of an architect should also not limited in the red lines.

墨尔本大学建筑学专业录取的刘宇恒设计作品（二）

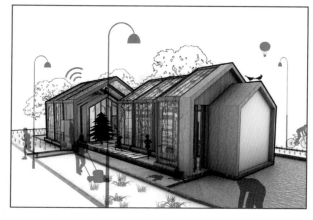
SUMMER

WINTER

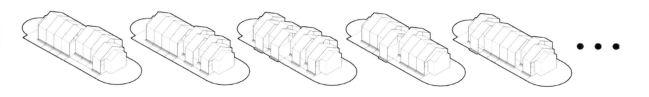

米兰理工大学建筑设计专业录取的耿直设计作品（一）

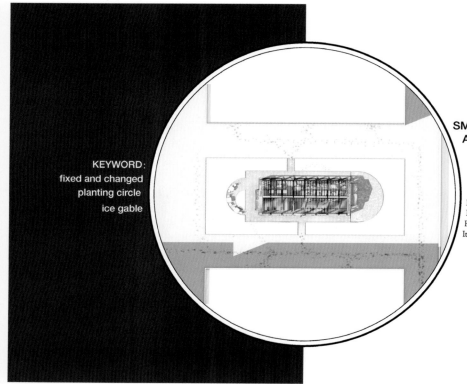

KEYWORD:
fixed and changed
planting circle
ice gable

SMALL COMMUNITY CENTER AND GREENHOUSE

(second semester in the fourth year of college, individual work)

Located in Mudanjiang City, Heilongjiang Province, the northernmost part of China. Five months of a year below 0°C. It is a parterre spared in autumn and winter

米兰理工大学建筑设计专业录取的耿直设计作品（二）

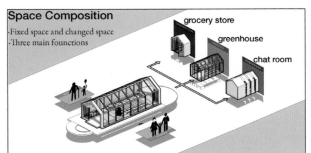
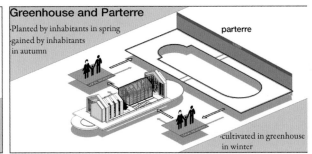
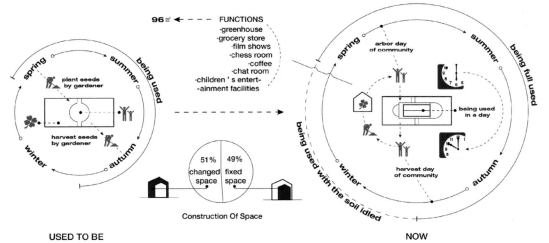

米兰理工大学建筑设计专业录取的耿直设计作品（三）

（二）境外求职的毕业生

除了申请境外院校的申请者之外，留学作品集的第二大受众群体就是打算在境外求职的毕业生。这部分同学一般已经有了一年以上的境外学习经历，对留学作品集的内容和形式有了比较多的了解。境外求职的作品集往往是在原有境内作品集基础上，加入境外学习阶段的设计作品，按高标准的学术作品集要求制作的求职作品集。我们前面已经讲述过留学作品集和境内工作作品集之间的几点主要区别。那么境外求职的作品集是否和境内作品集有很多相似之处呢？答案是肯定的，境外求职作品集也是应对职位申请者筛选的重要证明材料，不过它和留学作品集又有更多的相似性。我们可以视作是在留学作品集基础上发展起来的一类作品集，因为境外的设计事务所和公司往往更学术化和注重作品的结构逻辑和思辨性，特别是在境外刚毕业的本科生或研究生，在没有丰富的实习经历前提下，求职作品集的主要内容还是以在校期间的学术作品为主。

（三）工作室设计师

作为比较特殊而新颖的一类设计师，事务所的设计师们往往也需要优秀的作品集来作为设计理念的宣传和推广材料。在欧洲，许多优秀事务所往往都保留着在网站展示学术作品集的特征，比如著名的丹麦 BIG Bjarke Ingels Group 公司、荷兰的 MVRDV 公司、英国的 Foster + Partners 等，熟悉这些公司网站的同学们一定对他们精致独特的分析图、设计理念的阐述以及草图的绘制过目难忘。这与境内许多设计院的作品展示有着极大的区别。浏览境外建筑师的网站往往更容易让人想起学校评图的场景，故事化、情景化和学术化氛围更浓。

第二节

通过前面的章节想必大家对留学作品集本身以及和其他作品集之间的区别有了一定的认知。接下来我们将重点介绍作品集在申请中的注意事项。

一、设计类留学申请的特殊性

和其他专业不同，设计类专业留学可以说是最特殊的一类，主要特殊就是需要展示优秀的设计能力和天赋，从申请境外的设计类本科到硕士、博士，作品集都是最重要的材料之一。相比物理、化学、数学等理科专业，历史、经济等人文社科专业，道路、桥梁、电子、机械等工科专业，设计类专业可以说是最特殊的一类专业。特殊就特殊在专业的归属性和录取评审标准的特殊性。

以归属性来说，相比其他明确属性的专业，设计类专业的归属在境外不同国家或地区不同院校也有很大不同。比如建筑学专业，在有的学校可以属于社科Social science（香港大学），在有的学校属于工科Engineering（欧美院校将建筑学归于环境工程学院或建筑与土木学院，德国的建筑师毕业大多被授予Ingenieur工程师称号），在一些学校又隶属于艺术类专业Art（比如传统老牌的英国格拉斯哥艺术学院、英国皇家艺术学院等），还有一些学校将建筑学和相近专业专门独立出来（比如哈佛GSD设计学院、UCL伦敦巴来特建筑学院等）。不过面对这么多眼花缭乱的分类，

同学们不用担心，建筑学是与众不同的设计专业，是艺术与技术结合、通才与专才教育的佐证。不管在哪个国家或地区，哪个院系下的建筑学专业毕业，只要是教育部认可的正规院校（世界名校就更不用说了），都会等同于我国的建筑学专业教育，并且在就业和个人发展层面往往有非常大的优势。不同的归属性直接会影响申请材料的准备，特别是作品集的准备。比如哈佛GSD设计学院和丹麦皇家建筑与艺术学院的作品集就更应注重艺术和纯创造力和对空间认知的表达，而德国慕尼黑工大、荷兰埃因霍芬理工大学、代尔夫特理工大学建筑学的硕士入学作品集则需不同程度上兼顾艺术与技术、建筑空间、建筑结构和建筑技术。因此同学们在申请院校的选择和作品集的制作之前需要给自己良好的定位。这样才能有的放矢，把宝贵的准备时间和精力最有效率和针对性的利用起来，做出好的作品集。

下面列举一些最知名的设计类院校的作品集详细要求：

（1）瑞典皇家理工学院作品集要求。（https://www.kth.se/en/studies/master/architecture/entry-requirements-1.48053）

（2）哈佛大学设计学院GSD作品集要求。（https://www.gsd.harvard.edu/admissions/application-instructions/）

（3）代尔夫特理工大学作品集要求。（https://www.tudelft.nl/en/education/programmes/masters/architecture-urbanism-and-building-sciences/msc-architecture-urbanism-and-building-sciences/admission-and-application/non-dutch-bsc-degree/）

除了归属性上的特征之外，录取评审标准的特殊性也是设计类专业与众不同的特点之一。其他大多数专业的录取最主要的审核标准是本科的GPA，专业排名，本科院校的知名度，语言成绩，申请文书的优秀程度等。而设计专业则可以说是独尊作品集，作品集在录取评定中往往占60%左右的分量。其次才是申

请文书，平均成绩绩点，本科院校的知名度等。许多世界名校对设计专业的平均分（GPA），语言成绩都不做非常苛刻的要求，这一方面体现了设计类教育更开放注重设计本质的特点，另一方面也让许多本科院校不那么占优势的学生能有机会进入到世界一流的大学就读。比如英国的 UCL（全球排名前 10 名），澳大利亚顶级学府墨尔本大学（全球前 30 名）都对国内许多非 211 大学的学子敞开大门。意大利最优秀的理工大学米兰理工大学（同时也是普利策奖大师伦佐皮亚诺和阿尔多罗西的母校），近几年对建筑学专业的平均分录取要求也仅仅在 75 分左右，并且给非 211 院校的优秀中国学生授予白金奖学金，拥有英国最大建筑学院的名校谢菲尔德大学也曾为雅思小分不达标的设计类学生专门开绿灯发放无条件录取并出具协助签证材料。种种现象都说明设计类院校是非常灵活开放的，但是这所有的一切都需要有一个强有力的前提：非常优秀出众的设计作品集。这里并不是说学设计的同学就可以不注重 GPA，不好好学语言，这些也非常重要，只要是涉及评定的项目，都非常重要，但是最重要的是设计作品集。一份好的作品集可以在一定程度上弥补学校背景甚至平均分不那么优秀的短板。如果对方评审教授觉得申请者的设计作品足够优秀，那么伯乐识千里马的故事就很可能一再上演。

二、专业申请者的困惑：孰轻孰重？语言，文书，作品集？

说到这里很多同学就会有疑问，这么说既然作品集是最重要的，那么其他材料应该如何准备和分配精力呢？有哪些材料是需要非常重视的，语言、作品集、文书应该如何安排呢？

（一）作品集的重要性

凡事预则立，不预则废。留学申请是一个非常重大的系统工程，有合理的进度规划和时间安排才能达到最好的效果。在整个规划中，作品集作为最主要的申请材料，应当放在最首要的位置。基础好的同学可以更上一层楼，通过制作一流的作品集冲刺世界顶级的院校，基础相对较薄弱的同学可以集中优势兵力先突破作品集，集中力量办大事，先把最难最花精力的作品集做好再进行其他材料的准备

I was inspired by a special chemical material---zeolite. I studied zeolite when I was in a chemical lab. And when I used Electron Microscope to observe it, I was shocked by its beautiful and neat structure. Zeolite is used for aborbing or filtrating things, given its special micro-structure. It has lots of micro-structure whose characters are the same---holes, rooms, the ability of extention. Inspired by its structure, I believe this chemical structure could be used in architecture.

zeolite image through Electron Microscope

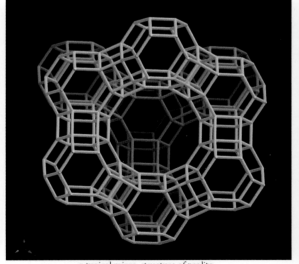

a typical micro-structure of zeolite

美国USC March1录取的黎子阳转专业设计作品（一）

和制作。在这里我们建议同学们无论是申请哪个国家或地区的院校，都尽量提前做好规划，一步一个脚印脚踏实地，把申请任务细分成几个阶段，每个阶段集中精力解决一两个主要矛盾。而作品集往往就是在第一个阶段最先需要解决，也是工程量最大的主要任务。

（二）文书与作品集协同重要性

在上面的小节里我们提到了作品集的重要性，作为首要的申请材料，需要安排最多的时间和精力来集中突破。但是除了作品集，文书也是设计类专业非常重要的材料，这往往是很多同学容易忽略的，文书在某种程度上甚至要超出 GPA 和院校背景在申请审核中的分量。因为文书同样是作为故事性叙述的重要材料，申请文书根据申请国家，院校不同有所区别，主要包括：动机信、个人陈述、简历、推荐信、学习计划、奖学金申请信，一些院校还需要提供证明专业英文能力的小论文（writing example）。文书的种类和具体要求因学校而异，最重要的四项是动机信、个人陈述、简历、推荐信。澳大利亚和英国一些院校会将动机信和个人陈述综合到一起，北美院校一般将这两项分开。而一些对理论素养要求较高的院校会在审核的第二阶段要求提交 writing example（比如 UCL 的建筑及城市历史专业）。很多同学容易犯的误区是重视作品集而忽略文书，殊不知文书在评审过程中同样占有非常重的分量。和申请很多其他专业不同，设计专业的文书需要体现申请者的设计思想和成长历程，对设计的理解和具体专业方向的兴趣。以建筑学留学文书为例，建筑设计的流派，各位大师的设计理论，现行的建筑设计思潮，前沿的建筑实践活动，所申请院校的风格这些都需要在文书中翔实地体现出来。因为设计行业是一个高度专业化的行业，专业的文书和泛泛的文书在审核委员会眼中立见高下。试想，如果文书中阐述设计观点时，如果连理论名词，流派特征，院校的研究方向都不能清晰表达，只是泛泛阐述自己的兴趣和论述普通的申请理由，一定会让申请大打折扣。所以，在文书上也需要做足功夫。最合适的方式是文书与作品集及相互配合，形成一个体系，切不可泛泛而谈，而应该有针对性，有自己的特点，和作品集的风格相匹配。作品集是直观的表达，以图示语言的形式来呈现专

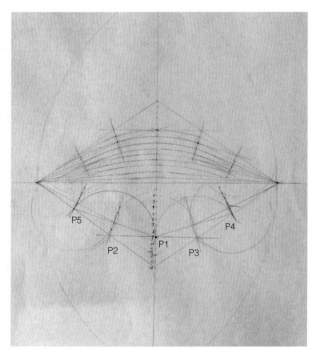

Drawing of Waving Mechanism

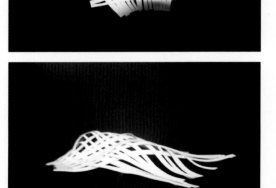

Waving Study Model

美国USC March1录取的黎子阳转专业设计作品（二）

业设计能力与思想，而文书则从专业的角度娓娓道来，来补充说明申请的动机和个人的优势与学习计划。只有文书和作品集相互配合才能发挥最佳组合作用，同时给评审委员会留下较深刻的印象。

三、转专业申请者的终极兵器：打开名校大门的金钥匙

在众多申请者中，不乏有一些对设计专业充满极大热情的转专业的同学。我们见过许多本科生由于种种原因不能就读于自己理想的设计专业的同学，但是他们心中对建筑或是其他设计专业还一直念念不忘。对这些同学而言，留学转专业申请建筑或是其他相关设计专业则成为努力实现自己理想的极其重要的一个转折点。美国与澳大利亚许多的知名院校可以接受本科非设计专业的学生转专业申请，但与此同时，由于是转专业申请，本科的院校、GPA、原专业老师的推荐信都在评审过程中显得不那么重要，作品集成为了几乎唯一的衡量标准。以建筑学接受转专业学生申请为例，美国大多数顶级建筑院校都接受转专业学生直接申请建筑学硕士，称作 March1（多为 3 年制，有别于有专业建筑学本科学生申请的 2 年制 March2）。在 March1 的申请中，多数院校都明确指出作品集是重要的申请衡量材料，是体现申请者是否具有足够的创造力，是否拥有符合录取标准的视觉表达与空间理解能力，三维空间构思能力以及优秀的设计潜质的重要佐证材料。以美国前世德 USC 南加州大学建筑系 March1 作品集要求为例，转专业作品集需要包括：①设计者独立设计绘制的设计图纸；②过往参与的建筑设计作品的模型照片；③参与过的实际工程相关图纸或者其他能表达设计者创意的设计作品。作品集要求 10 至 25 页，在网站 https://uscarch.slideroom.com/ 上传，总大小不超过 25M，可以包括其他媒体形式的链接，如 Youtube 或者 Vimeo 视频链接，但是不接受任何形式的 CD/DVD 以及打印的纸质作品集。(https://arch.usc.edu/apply/graduate-admission-0)

Waving and Aggregation

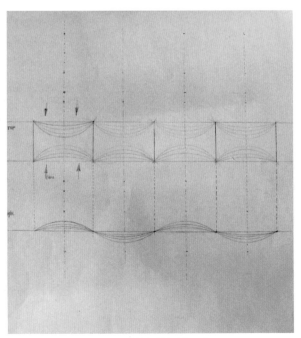

Mechanism of Making Unit of Paper Bar

Twisting paper bar by special mechanism can make one paper bar into sevral special units. And by waving these paper bars with units, the paper bar can be formed as a tight and stable structure which looks like a net and can be twisted into several shapes just as the bsmboo waves

This structure is very stable because when two unit combine, the pressure of these two unit can be counteracted as a result of the special structure.

美国USC March1录取的黎子阳转专业设计作品（三）

The structure also has some aesthetic features.

The structure has a neat topographical feature. The units aggregate, repeat and sometimes has changes

When sunlight go through it, fantastic effect of light and shadow occurs.

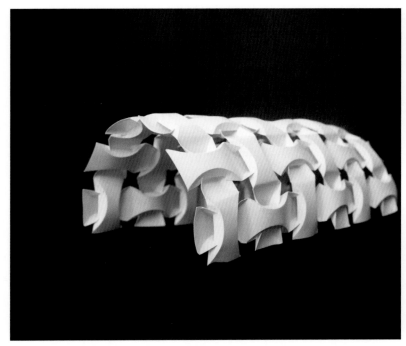

美国USC March1录取的黎子阳转专业设计作品（四）

由于作品集是转专业学生申请设计类专业的终极兵器，所以更加应当重视，而且制作安排应该注意和非转专业同学作品集区分开来，在展现一定的专业技能的同时更应体现自身的创造力、设计的特色以及可塑性潜力。我们将作品集称为转专业学生打开名校设计专业大门的唯一金钥匙毫不为过。

什么是好的留学作品集
The Characters Of A Good Application Portfolio

第三节

介绍了这么多，很多同学一定非常关心，作品集如此重要，那么怎样的作品集才能称之为一本好的作品集，或者达到什么样标准的作品集才能让评委老师眼前一亮，对自己的作品青睐有加呢？接下来我们就将为大家一一解读怎样才能制作出一本国际优秀水平的作品集。

一、拨开重重迷雾，申请者的误区解读

要做一本好的作品集首先就是要避免许多申请者常犯的错误，避开常见的作品集误区，不要让原本很有希望的设计被一些低级的错误手法或者认识上的偏差带偏了。我们总结许多同学在申请初期常犯的错误以及之前学习过程中养成的不好习惯，主要分为以下几点：

（一）误区一：过分重视图面效果

在当今的读图时代，图面效果十分重要。一本好的作品集往往都有优秀甚至完美的图面表达。因此许多同学往往认为只要有了好的效果图和分析图以及惊艳的排版就可以棋高一着。其实不然，这也是许多同学容易走上的歧途之一。好的图面固然重要，也是境外院校非常希望见到的，但是，好的图面不等于炫酷的效果图和分析图组合。炫酷的图和高质量的图有着本质区别，在我国的竞赛，我国的许多设计院商业投标过程中，炫酷的效果图以及天花乱坠的分析图往往是受到许多甲方以及设计师推崇的。而实际上，留学作品集是非常学术的作品。我们可以这么说，天下武功，唯设计不破。有了好的设计，就不怕出不来好的作品集。学术的作品集最讲究的是图纸表达的学术内涵，设计思想和通过图纸展现的设计功底和知识底蕴。不需要炫酷耀眼的PS和渲染表达，每一条线都恰到好处，不多不少，增之一分则太过，少之一分则单薄。这是大多数国际知名院校所愿意看到的作品集。一定要牢记，作品集首先需要展示的是申请者的设计能力，对空间的表达和思辨能力，切不可变成效果图和分析图的堆砌。申请者要告诉对方的是：我是一个设计师，来仔细看我的设计吧，平面图、剖透图、空间转化等，而不是说，我是一个画图匠，快看我的效果图。再一次强调，建议留学作品集里不要出现商业化的效果图和超现实的竞赛效果图，评审团都是经验非常丰富、眼睛雪亮的教授和设计师，空洞没有足够内涵支撑的图面和耍酷的后果往往就是直接被否决掉。这样很多同学又会问，如果不追求炫酷的图面那么什么样的图才是好的学术图呢？正如一千个观众眼中有一千个哈姆莱特，我们不建议给出一个单一的标准和模板，事实上也没有什么图纸表达能够可以说是天下第一。北欧高冷风格，日系淡雅细腻风格，德国哲学系风格，北美叙事风格和南欧插画风格都可以做出非常优秀的图纸。他们有着共同的特点：清晰，直截了当，真诚朴实，简单易懂。我们举一个例子，如果一个标识你在3秒钟还不能记住，那么它就不是一个好的图示表达。请大家闭上眼睛回想苹果手机的标识和家乐福超市的标识。相信大多数同学都能准确地回忆出苹果的标识，可能有同学会记错被咬掉的一块是在左边还是在右边，但是家乐福的标识大多数同学不能那么清晰地记起是怎么样的。这就是表达清晰度简单易懂的意义。如果一张图太过冗繁或者画蛇添足缺失了重点和清晰性，当评委老师看不懂的时候他们一定不会觉得是他们的问题。因此，大家应当把有限的时间和精力更多地用于方案的设计和图纸的精准表达上。世界上许多著名建筑师的优秀作品都是以朴实，细

腻的风格，用最简单的建筑语言来直达人心的。比如瑞士建筑师彼得·卒姆托，美籍华人建筑师贝聿铭，以及日本建筑师安藤忠雄和意大利建筑师卡洛斯卡帕等。

（二）误区二：作品集以量取胜

第二个常犯的错误是作品集以量取胜，很多同学容易认为给学校展示的作品应该是越多越好，这样可以显示自己做的设计经验丰富，种类繁多。殊不知，境外院校最青睐的作品集是少而精，优而全。少而精指的是作品的数量不宜过多，一般以4个作品左右为最佳，3个显得单薄，5个略显累赘（特殊情况下比如有较长的工作经历除外）。优而全指的是作品质量要出色，涵盖内容要全面。以建筑学作品集为例，建议涵盖不同的由小到大的建筑类型的作品。学术的，竞赛的都可以包含在内。我们见过有的同学在作品集里放了超过10个设计作品，冗长的作品集特别容易让读者感到审美疲劳和缺乏耐心。许多学校对作品集的大小页数都有明确的规定。以澳大利亚最高学府墨尔本大学建筑系为例，建筑作品集要求不超过15页，每个作品约三四页左右。学校虽没有指明一定要几个作品，但是包括封面目录一共15页的作品集意味着最佳的作品数量就是4个。英国与荷兰的许多院校在申请信息页面明确指出，作品集是申请者最佳的作品集合而不是全部的作品集合。这样，集中精力来做好4个左右的设计作品远远胜过仓促组合的一堆设计作品。

（三）误区三：作品集均为高年级作品

第三，许多申请同学容易犯的错误是作品集里均为高年级的设计作品。这个误区不是非常严重，但是在能够避免的情况下希望同学们能尽量避免。的确许多同学会觉得低年级的作品多为手绘作品，设计作品还没有入门，很多方面都不够完美，因此不太想放到作品集之中。但实际上，在作品集中适当放一到两个低年级或者大三的作品是很合适的。不建议全放大四、大五的设计作品，因为许多院校希望看到申请者设计思维的进步，对设计理解的逐步提升。那么，如何来解决低年级很多设计作品不够成熟，图纸不够完美的问题呢？最好的方式就是在低年级设计作品的基础上尽心深入地修改或者重新设计。这样既可以保留最初学习设计时许多富有创造力的想法，又可以用比较成熟的设计语言和空间营造手法来创造出优秀的设计作品。此外，竞赛的设计作品也是很推荐放入留学作品集当中的。

（四）误区四：实际案例作品过多

一些学生的作品集中放入了过半的实际工程案例，甚至有的同学整个作品集全都是由实习和工作的作品组成的。这也是作品集中容易出现的误区。和工作求职作品集不同，留学申请作品集是以在校学术作品为主，院校的审核也多是以学术方案的标准来鉴定。由于工作作品有很强的特殊性，比如商业化强，甲方对设计的影响较大；工程技术性强，理念和空间设计相对弱化；作品均为团队设计，没有充分体现申请者单一设计作品的表达等问题。因此我们不建议放过多的工作实习案例。应当还是以学校设计作品为主，可以辅助添加实际工程和其他体现设计能力和创造力的设计作品。

二、优秀作品集必备的要素

以上是我们在制作作品集过程中应当注意的误区，应当避免的错误，下面我们给大家讲解一下优秀作品集制作过程中应该做的要点。

（一）清晰的图面

清晰的图面是优秀作品集的第一要素。留学作品集忌讳满满的图面，天花乱坠的表达。简洁大方的排版和清晰的图纸能给人留下优秀的第一印象。作品集每一页不需要展示太多的内容，不要试图将过多的信息加在一张图上，每页都有核心内容和辅助图示。这一点非常类似路易斯·康说的服务空间和被服务空间：每一页都有服务图示和被服务核心图示。每张图说清楚一件事情，能用一张图说清楚的事情绝不用两张图来解释，能用图示说明的就不用文字说明。

除去图面排版和内容的清晰之外，图面色调和风格的统一也是非常必要的一点。和谐的色调和图面画风可以让表达更具连续性和易读性。

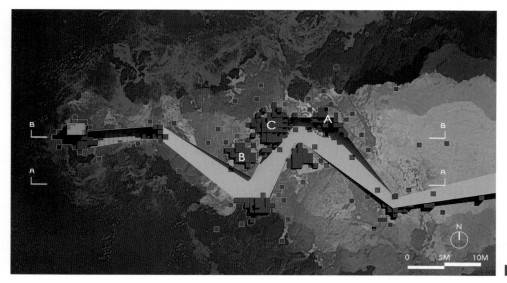

MASTER PLAN

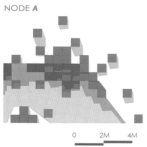
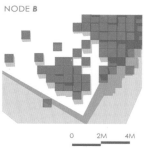
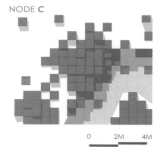

NODES PLAN

米兰理工大学建筑学专业录取的白金奖获得者李炫静设计作品（一）

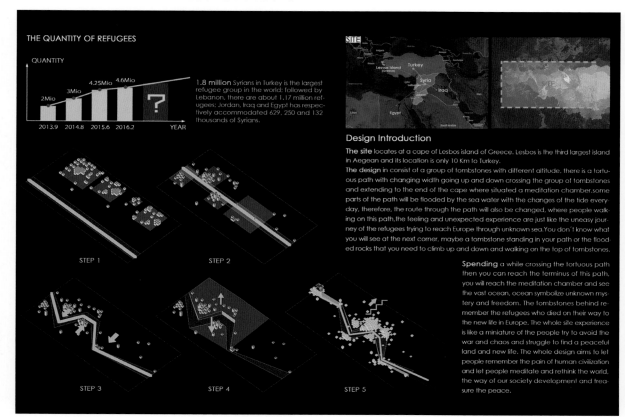

THE QUANTITY OF REFUGEES

1.8 million Syrians in Turkey is the largest refugee group in the world; followed by Lebanon, there are about 1.17 million refugees; Jordan, Iraq and Egypt has respectively accommodated 629, 250 and 132 thousands of Syrians.

Design Introduction

The site locates at a cape of Lesbos island of Greece. Lesbos is the third largest island in Aegean and its location is only 10 Km to Turkey.

The design in consist of a group of tombstones with different altitude, there is a tortuous path with changing width going up and down crossing the group of tombstones and extending to the end of the cape where situated a meditation chamber. some parts of the path will be flooded by the sea water with the changes of the tide everyday, therefore, the route through the path will also be changed, where people walking on this path, the feeling and unexpected experience are just like the uneasy journey of the refugees trying to reach Europe through unknown sea. You don't know what you will see at the next corner, maybe a tombstone standing in your path or the flooded rocks that you need to climb up and down and walking on the top of tombstones.

Spending a while crossing the tortuous path then you can reach the terminus of this path, you will reach the meditation chamber and see the vast ocean, ocean symbolize unknown mystery and freedom. The tombstones behind remember the refugees who died on their way to the new life in Europe. The whole site experience is like a miniature of the people try to avoid the war and chaos and struggle to find a peaceful land and new life. The whole design aims to let people remember the pain of human civilization and let people meditate and rethink the world, the way of our society development and treasure the peace.

米兰理工大学建筑学专业录取的白金奖获得者李炫静设计作品（二）

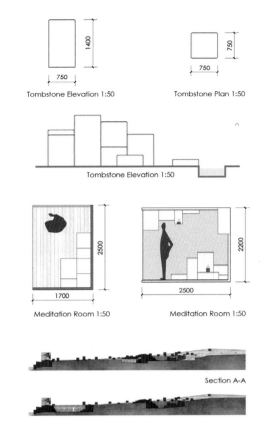

米兰理工大学建筑学专业录取的白金奖获得者李炫静设计作品（三）

Site History
This is an unbuilt site of the branch of Dongbei University in Qinhuangdao. The picture shows the relationship of the new site and the old one.

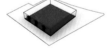

Rise
The volumn rises according to boundaries and take full use of the available space.

Seperation
Divide the volumn into 4 parts, an offical building for teachers and staffs, three teaching buildings for students, respectively.

Revolve
Revolve the offical building to be oriented to south, the remainder three buildings are oriented to the direction of the street. These two axis are oriented to the surrounding city texture.

Connection
The 4 volumns were all isolated originally, but the 3 buildings used by students should be taken full advantage of if connected by skyways. The skyways also planted by virescence, so the students in the 3 buildings can touch the virescence easily.

Landscape
The axis of landscape starts from the "Chensi Plaza", extend along with the "Xi Road" and finally arrives at the destination (the central plaza of the new campus.) the landscape does make the relationship between the old and the new be more close.

Style and colour
This area has a distinct feature: a certain sum of the buildings are painted red, there are also some red constructions in the old campus, so the main colour of the new is red too.

Final Solution
The 4 buildings seem to be intergration through the axis, colour, texture and the skyways, also, the landscape will be found everywhere.

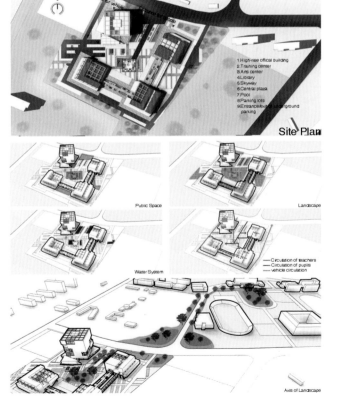

米兰理工大学建筑设计及历史专业录取的蔡笑革设计作品（一）

19

Overview

The size of Northeastern University at Qinhuangdao is not very large, it covers an area of 470,000 m² and has more than 800 teachers and staffs. Now, the small size has restricted the the number of pupils, which hinders the development of the university. The authority decide to extend the size of the university.

Site Location

The new site locates in the southwest to the main campus. There is just one road seperating the these two, so it is very convenient for pupils and teachers to study and work.

Analysis Of Main Campus

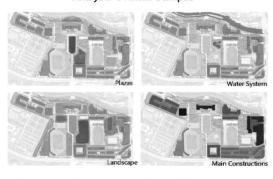

Site Analysis

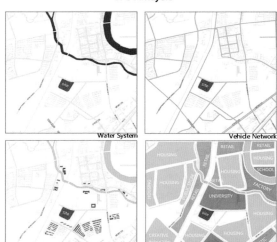

Surrounding Environment

There are a number of retails and residential areas surrounding the site, which can meet the requirements of the students and teachers.
Plenty of surrounding constructions are red, which has fromed a style of this area.

米兰理工大学建筑设计及历史专业录取的蔡笑革设计作品（二）

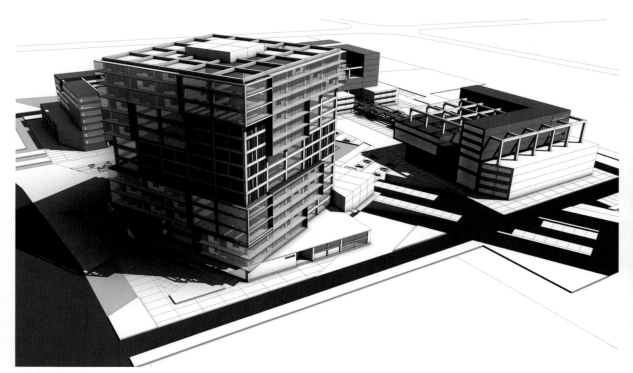

Stadio course assignment
4th Year Individual Work
Time Period: 8 Weeks
Location: Qinhuangdao, Hebei Province, China
Site area: 70818 m²

NEW CAMPUS OF NORTHEASTERN UNIVERSITY AT QINHUANGDAO

米兰理工大学建筑设计及历史专业录取的蔡笑革设计作品（三）

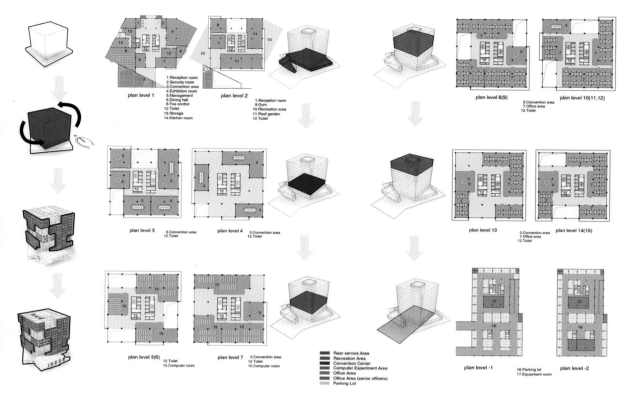

米兰理工大学建筑设计及历史专业录取的蔡笑革设计作品（四）

（二）强大的逻辑性

逻辑性是作品集非常重要的一点。好的作品集就像一本引人入胜的故事书，读者会手不释卷，很自然地从一章过渡到下一章。对作品集而言，如何科学地组织图面和表达顺序非常重要。每一张图所包含的内容都需要慎重选择，第一页要表达什么理念，第二页计划详细阐述什么问题，页与页之间如何顺畅连接，这些都是需要申请者认真规划的。好的逻辑结构会让读者自如的从一张图过渡到下一张图，不用提示就能清楚地知道每一页是从上往下浏览还是从左往右浏览。看完一页自然会有对下文的预期，而下一页的内容正好能衔接上。每一页都有自己的核心重点，图文并茂，相辅相成。而如果逻辑思路不清晰则很可能造成很混乱的效果，就像故事叙述者前后颠倒或把情节混淆在一起，读者读得一知半解，没有头绪就很容易失去兴趣，对作品的印象大打折扣。

因此作品集的逻辑结构是组织作品集表达语言的最重要大纲，有了好的大纲才能把细节有序的组织好，效果图、分析图、技术图、文字相互组合。

（三）独特的创造性

创造性是众多院校衡量作品集好坏的一个重要评分点，许多学校都明确指出 creativity（创造性）是申请审核的要素。下图是墨尔本大学建筑学专业对作品集的详细阐述，第一句话就提到了作品集是充满创造力的艺术，是用来体现申请者的技艺和想象力的。

创造性可以从很多方面来体现。以建筑设计为例，造型、空间、结构、材料等各个方面都可以做出很多有创造性的设计：新颖而符合结构和空间使用的造型设计；巧妙的结构使用和空间需求的融合；传统建筑语言的再诠释和再定义；材料的创意使用实现新的功能；对地形和基地的合理利用；变限制条件为有利因素等。比如韩国梨花女子大学覆土校园建筑的设计体现的是建筑与场地融合的创意，贝聿铭美国华盛顿美术馆东馆的设计体现的建筑空间和造型以及轴线秩序的统一创意，安藤忠雄

> "The portfolio is a creative act, showing your skills and imagination, but it is also an act of communication and a tool for self-promotion. Demonstrate originality and inventiveness, but also accept the restrictions and conventions of professionalism, and show that you can get your ideas across in terms that working architects and designers can understand."
> Harold Linton

墨尔本大学建筑学作品集官方介绍

的本福寺水御堂荷花池的设计是禅意诗歌的语言建筑物化表达的意境创意，王澍的宁波美术馆设计是传统建筑材料在新时期的再表达和再诠释。

（四）落地的技术性

一本好的作品集离不开翔实的细节描绘。重设计而轻技术是许多的申请同学特别容易犯的错误，特别是建筑设计专业，许多同学的作品表达在空间上，意境上都非常不错，但却缺乏相应的技术表达。建筑学是艺术与技术的结合。设计作品中应当适当的表达出对建筑结构、建筑物理、建筑设备等技术层面的知识储备和应用。许多学校把相应的技术表达作为作品集的重要评分标准之一。比如荷兰的代尔夫特理工大学在作品集的要求中就明确规定建筑技术是评分的依据之———"Applicant demonstrates appropriate knowledge of applied building technology... All criteria must be met. Failing on one of them can lead to a rejection" 任何一项的缺失都可能导致申请被拒绝。所以同学们在申请的时候尽量要全面的展示自己的知识结构体系，在合适的设计中体现自己对技术方面的踏实基础是非常必要的。

（五）人文、地域性的关怀

国际一流作品集往往都具有相似的特质——大巧若拙，质朴归真。不光是作品集，纵观不同国家不同大师们的作品，许多的大家推崇备至的作品都是非常质朴的，没有天花乱坠的吊饰，没有华而不实的手法，更多的是对空间、材质、建筑结构，对建筑本身和建筑使用者的认真思考。其中很常见的就是作品对历史、文脉、环境以及地域特征的关怀和对话，设计上升到一定的高度都离不开对设计本身以及容纳设计环境的再思考，就像我们在夜深人静的时候会自己问自己，我是谁，我从哪里来，要到哪里去。因此，好的作品集也离不开对人文和地域性的思考以及积极表达，而这个层级也是最能提升作品集层次，体现设计者创造力，功底以及对设计本质认识的层级。

第二章
留学作品集的准备
Preparation Of Portfolio

留学作品集的准备周期和时间安排
Preparation Period And Time Schedule

第一节

一、准备周期

作品集的准备周期因人而异，也根据目标院校的特点和要求有一定区别。以建筑/规划/景观专业毕业年级学生为例。中等偏上水平的学生一般准备周期需要5到6个月（每周连续投入作品集制作，同时学校的课程并行）。如果是已经毕业的学生可以全职做作品集，基础较好的学生准备周期在3个月左右。如果基础较为薄弱的学生则需要的时间更长，加强软件、加强设计理念和基础知识的培训都会占据很多的时间。转专业的学生制作作品集则需6到8个月。所以建议同学们提前做好准备，安排好充裕的时间来准备作品集。由于大多数国家的申请开放时间在毕业前6到12个月，因此最佳的作品集准备时间是大四下学期开学到大五上学期（4年制则为大三下学期到大四上学期）。

二、准备时间安排

准备时间的安排也有很多的讲究，确定好了申请的院校之后，就需要合理的规划准备时间，一份科学合理的时间安排表是申请成功的重要基础。影响时间安排计划的几大因素如下：

（1）申请目标院校的接受申请开放时间。
（2）申请目标院校的申请截止时间。
（3）作品集的基础程度。
（4）语言成绩递交的时间以及语言准备的基础情况。
（5）学校的课业强度。
（6）其他个人重要事情安排。

首先是目标院校的接受申请开放时间。

先说主流的秋季入学申请（每年9至10月份入学），澳大利亚提前一年接受申请，每

September 1st, 2016	Opening of the online application
November 24th, 2016	Closure of the online application/last day to upload and provide required documents online and by post
February 10th, 2017*	Last day to provide your language certificate (to be considered for merit based scholarships)
October 2016 - February 2017	Results notification
February 28th, 2017	Deadline for accepting the admission
March - July 2017 **	Contact Italian Diplomatic Representatives for documents legalization and Visa procedure
September 2017	Enrollment and Beginning of classes

米兰理工大学9月入学申请时间表

MSc programmes start fall semester

1 October 2017
Start of the application procedure for the MSc programmes starting September 2018.

1 April 2017
Application deadline (31 March 2017, 23:59 CET)
Advice for EU/EFTA nationals: upload proof of English language proficiency to make sure your enrolment and access to services can be arranged on time.

1 June 2017
Application deadline for European post-master in Urbanism (EMU) and the Berlage Post-Master of Science in Architecture and Urban Design for non-European students. For Non-EU residents, we highly recommend applying before 15 May 2017, in order to have enough time to arrange VISA and housing.

1 July 2017 (30 June 2017, 23:59 CET)
Confirmation Statement & full payment deadline
EU/EFTA nationals: upload proof of English language proficiency.

4 September 2017
Start of the MSc programmes

代尔夫特理工大学9月入学申请时间

年7月下旬接受第二年7月入学的申请；意大利9月初开始接受第二年9月入学的申请；英国每年9月中旬接受来第二年9月入学的申请；荷兰10月初开始接受第二年9月入学的申请；北欧以及北美院校11月底至12月开始接受第二年9月入学的申请。

部分国家的学校开设有春季入学（每年2月至4月入学），澳大利亚每年7月下旬接受第二年3月和第三年3月的入学申请；意大利每年5月中旬接受第二年2月入学的申请。

部分美国院校有春季入学，但不一定在网站上明确注明，需要提前和学校教学秘书联系确定是否招生以及具体的申请时间。

申请院校的截止日期（application deadline）是决定作品集和其他准备工作的最重要的节点。英国、澳大利亚的截止日期很晚（一般一直到入学前）但没有太多的实际意义，因为这两个国家都是先申请先录取，入学前4到5个月名校录取名额基本已满，意大利、荷兰、德国也是先申请先录取，所以建议一开放申请就投递，赶在第一批申请投出去，这样录取名校概率最大，因为越到后期录取名额越少，申请竞争也越大。截止后统一处理的国家有美国、加拿大、丹麦、芬兰、瑞典、挪威，这些国家的申请截止日期都在12月底到1月初（丹麦个别院校在第二年3月初截止），因此申请北美和北欧的同学作品集一定要在11月底或12月初完成。

以大四下学期4月份开学开始准备作品集为例，周期为5个月，到9月初完成作品集。在制作作品集的同时，还需要分配精力做学校的课程设计和其他作业。最佳情况是将学校的高年级课程设计与作品集制作相结合，按照作品集的要求来做课程设计，这样在后期修改的时候可以省去大量的时间和精力。每周投入在作品集上的时间应保证15到20小时左右（建议每天2到3小时）。总共投入的有效工作时间约在400小时左右（基础薄弱的同学视情况需要增加工作时间甚至翻倍）。作品集的工作时间建议选择在时间相对完整，精力比较充沛的时间段，比如晚上和周末，这样同学们可以集中精力、步步为营地打好作品集的攻坚战。既然选择

May 25th, 2017	Opening of the online application
July 25th, 2017	Closure of the online application / last day to upload and provide required documents online and by post
September / October 2017	Results notification
October 10th, 2017 *	Last day to provide your language certificate (to be considered for merit based scholarships) and deadline for accepting the admission
October / November 2017 **	Contact Italian Diplomatic Representatives for documents legalization and Visa procedure
February 2018	Enrolment and Beginning of classes

米兰理工大学2018年2月入学申请时间表

了留学实现自己更高的人生梦想，追求更卓越的自己，那么风雨兼程的不懈努力和付出是必经之路。同学们在准备作品集的这几个月里一定要保持坚定的信念和斗志，相信所有的付出和努力都会换来丰硕的回报和充满希望的未来。

在作品集制作的安排上，基础好的同学可以建议先从难度最大的作品开始做，集中优势兵力攻克最难的部分，同时积累了经验后面的作品就会更加顺畅。基础比较薄弱的同学可以先易后难，将复杂的大设计留在后期，但这同时需要更严格的遵守时间计划表，留出充足的准备周期来完成全套的作品集。

三、作品集和语言考试的时间分配

大多数的同学都需要在准备作品集的同时准备语言考试。除了少部分小语种国家（德国、日本）之外，大多数同学都是准备英语语言考试来达到目标院校的申请要求。申请欧洲大陆、英国和澳大利亚的同学准备雅思即可，申请美国的同学需要准备托福和GRE，加拿大的院校GRE不是必选项。如何协调好语言学习的计划和作品集的计划，是一件至关重要的事情。对设计专业而言，语言成绩不是衡量入学的评分依据（达标即可，高分并不会增加显著的砝码）。而作品集则是录取委员会将认真审核评分的申请要素，因此我们建议在时间允许的情况下将更多的精力投入到作品集的制作中。但是语

言也必须达标，对很多英语基础不是很好的同学而言，语言学习的过程也不轻松。由于语言对词汇的基础要求很高，扎实的词汇功底会给高阶的考试冲刺准备带来非常大的好处，我们建议在准备作品集的同时可以学习语言的词汇。一套作品集准备下来语言的词汇基础也掌握得差不多了，然后集中准备雅思，托福或者GRE的应试训练，这样的效果往往都很好。把枯燥而需要日积月累的词汇基础学习和作品集准备结合起来，将语言基础的时间投入化整为零，一方面保证了基础学习的必要时间量，另一方面也可以给作品集制作做一定的调节。我们就是这么建议的，作品集做累了就通过学英语来调节！而作品集完成后，英语的应试训练可以在2到3个月左右集中完成，应试训练不宜做长时间的拉锯战。这样语言成绩可以在11月到12月初考出来（大多数同学需要考2次或2次以上），正好能满足申请时间的要求：英国和澳大利亚是先投递作品集和文书，拿到conditional offer然后补交英语成绩；意大利英语提交时间比作品集提交时间晚两个半月；北欧和北美作品集和语言成绩需要同时提交。

Timetable for IELTS Preparation Course 25H/20H per Week

	Monday	Tuesday	Wednesday	Thursday
9:00 – 10:00	Core Textbook	Core Textbook	Core Textbook	Core Textbook
10:00 – 11:00	Core Textbook	Core Textbook	Core Textbook	Core Textbook
11:00 – 11:15	Morning Break			
11:15 – 12:15	Workbook	Workbook	Workbook	Workbook
12:15 – 13:15	Lunch Break			
13:15 – 14:15	Skills Practice*	Skills Practice*	Skills Practice*	Skills Practice*
14:15 – 15:15	Intensive Skills Practice**	Intensive Skills Practice**	Intensive Skills Practice**	Intensive Skills Practice**

This timetable is an example and subject to change with or without notice.

*Skills Practice – Reading, Writing, Listening & Speaking
**Intensive Skills Practice – To reinforce each student's weaknesses

Mon – Thu – Students learn and practice Reading, Writing, Listening and Speaking skills required to successfully complete tasks in the IELTS Exam. This is achieved by working through a core textbook, doing activities from a workbook to practice all the necessary skills and by doing fortnightly practice tests.

25 hours/Week These blocks are only for 25 hours per week course.

Friday – Full Practice Test every fortnight.
A Progress Test and Revision of Units studied in the textbook every alternate Firday

	Friday
9:00 – 10:00	Full Practice Test
	Revision of Units
10:00 – 11:00	Full Practice Test
	Progress Test
11:00 – 11:15	Morning Break
11:15 – 12:15	Full Practice Test
	Revision of Progress Test
12:15 – 13.15	Lunch
13:15 – 14:15	Full Practice Test
	Supervised Self Study
13:15 – 15:15	Full Practice Test
	Friday Activities

CRICOS Provider Code: 01351B

（IELTS官方网站给出的准备雅思的时间安排表范例之25小时/周）
http://icqa.com.au/IELTSPreparationTimeTable.pdf
https://www.ets.org/s/toefl/pdf/toefl_student_test_prep_planner.pdf

第二节

一、软件水平要求

除了时间上的安排，另一个大家十分关心的问题就是作品集的软件要求。无论是基础比较好的同学还是对某些制图软件较为生疏的同学都总会问自己：我的软件水平够好吗，能不能达到境外留学作品集的要求，是否能适应境外留学的学习？第一个问题很多同学其实自己应该有一个清楚的答案，但是软件的水平可以不断提高。而设计师最重要的才能是思考模式和设计本身，软件是将设计细化以及可视化表达的工具。是否能达到境外留学作品集的要求，我们可以肯定地说，大多数的同学经过4年左右的认真学习，对软件的使用水平是完全可以胜任作品集的制作准备的，但是同样需要在准备中不断提高。因为涉及不同申请院校的风格，对图样表达的要求，同学们在准备作品集的同时需要有针对性地提高自己的软件技能，特别是在精细建模、分析图绘制以及渲染表达这几方面有针对性地系统加强。对于软件的水平能否适应境外留学的学习，我们给出的答案是：认真准备按高标准严要求完成申请作品集之后，申请者的软件水平是可以满足境外留学入学的软件使用要求的，正式开始境外留学的学习之后，还有很多要学的，这里面会涉及很多具体的建模和分析软件（不一定是绘图表达软件）。每个院校对软件的使用也有很大的区别，比如节能模拟软件、结构分析软件、风荷载计算和日照分析软件等，有的教授还会让同学掌握学校自己开发的软件，这些都是需要同学们去了境外留学之后在课堂上认真学习不断提高的。因此同学们在准备作品集过程中不需要这些非主流或者非常用技术流的软件，把常用的软件AutoCAD、Sketch Up、Photoshop、AI、Rhino等"武术基本功"融会贯通，练到炉火纯青就足够应付作品集的要求了。

（一）AutoCAD的水平要求

先说作品集对CAD的要求，CAD是所有设计师都通用的最朴实最基础的一种软件，我们用CAD主要制作的是作品集中的平面、剖

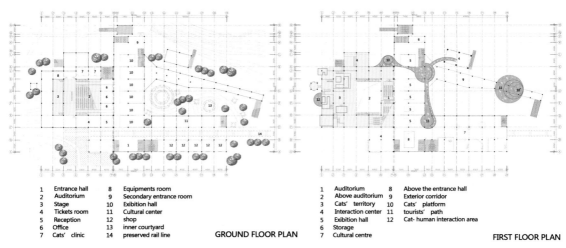

CAD绘制的设计平面图　设计者　瑞典皇家理工学院吴月

面和立面图（CAD 的功能及插件非常强大，我们用的可以说只是九层神功的第一层），CAD 和其他绘图软件比如 Adobe Illustrator、Photoshop 结合可以得到非常好的图面效果，既保留严谨的工程学特点，有能保证高品质的设计美学效果。

（二）Sketch Up 的水平要求

说到 Sketch Up 草图大师，大部分同学一定不会陌生，这一款非常好上手的软件伴随许多同学度过了无数难忘的绘图时光。Sketch Up 软件也几经易手，从 Last Software 到 Google Sketch Up 到如今的 Trimble 公司，几经波折，但始终是设计师们最喜爱的建模和作图软件之一。一般对 Sketch Up 软件的翻译是草图大师，这让许多同学认为 Sketch Up 就是做方案概念阶段用的软件，不精细，不适合做后期表达。其实这是一个很大的误区，Sketch Up 发展到如今，已经丝毫不逊于许多其他 3D 建模软件，它的人性化、易操作界面是最大的优势之一，仅仅在于曲面和不规则形体建模和参数化设计上还有待提高。无论是精细建模，还是渲染和辅助出图，Sketch Up 都是非常适合作为作品集准备的工具的。同时，Sketch Up 的 3D Warehouse 中有大量非常好的家具和组件模型，给设计师们提供很多的方便并节省了大量建模时间。如果同学们能够用好 Sketch Up，结合 Photoshop 以及其他软件，可以做出达到世界顶级名校录取要求的顶尖作品集。

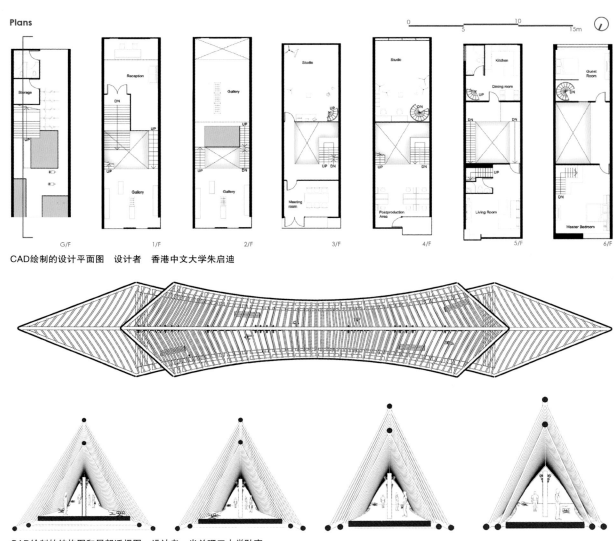

CAD 绘制的设计平面图　设计者　香港中文大学朱启迪

CAD 绘制的结构图和局部透视图　设计者　米兰理工大学耿直

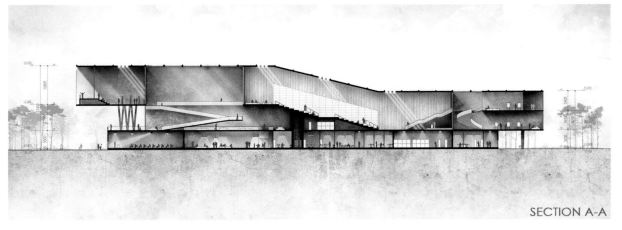

SECTION A-A

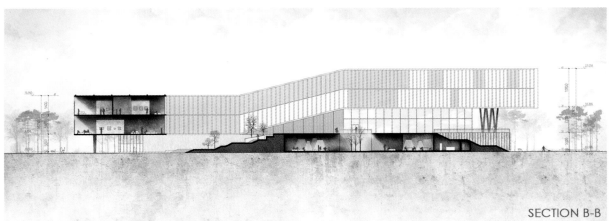

SECTION B-B

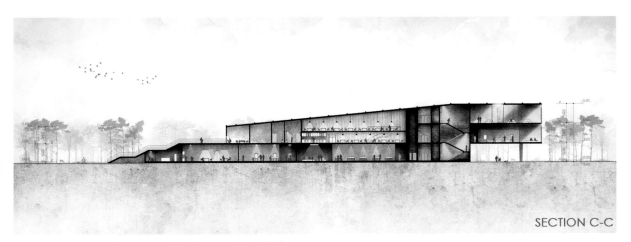

SECTION C-C

（CAD+Photoshop绘制的剖面图　设计者　米兰理工大学李炫静）

当然，这对软件的使用者也提出了更高的要求，既然 Sketch Up 可以作为作品集制作的主力软件之一，那么我们就不能再将其视为"草图工具"，而是一开始就需要养成良好的精细建模以及系统分组的建模习惯。对建模的熟练程度可以说是越高越好，在良好的模型基础上，配合 Vray for Sketch Up 可以做出初步的效果图底图。再辅助以 Photoshop 等修改软件，Sketch Up 可以支持从立面图、透视效果图、鸟瞰图、剖透图、节点透视等作品集中绝大多数图纸的出图需求。

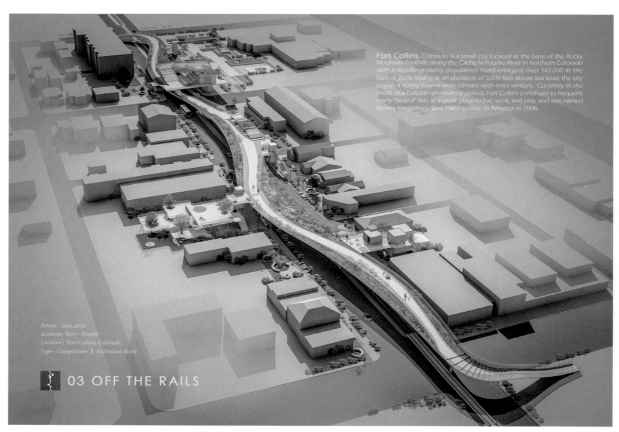

Sketch Up建模渲染+Photoshop制作的景观设计图　　设计者　李炫静

Sketch Up建模渲染+Photoshop制作的室内透视图　　设计者　李炫静

Sketch Up建模渲染+Photoshop制作的人视角度透视图　设计者　吴月

　　Sketch Up建模加渲染生成的效果图的特点是清新素雅，加上易于操作修改，使用得当可以大大提高作品集准备的效率。在制作分析图方面，Sketch Up也是灵巧好帮手。许多同学对丹麦BIG公司的分析图一定有着深刻印象，这些清新简洁的分析图能够十分明了地把设计理念和方案演化步骤表达出来。结合Adobe Illustrator、Sketch Up，在制作这一类分析图过程中也可以起到很好的支持作用。

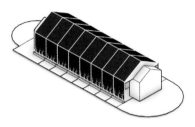

SPRING

Plant the creepers in the planting box

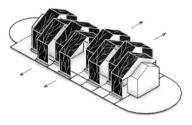

SUMMER

Open the skin to get better ventilation, and the grown-up creepers can replace the sun visor

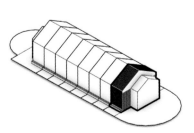

WINTER

Build the gable with bricks made of ice to against -30°C
The insolation of ice is much better than concrete especially wih air

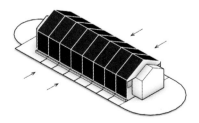

AUTUMN

Close the skin to keep warm from the cold wind. More sunlight is able to get into the house because of the wilt-ing of the creepers

Sketch Up建模+Adobe Illustrator制作的形体变化分析图　设计者　耿直

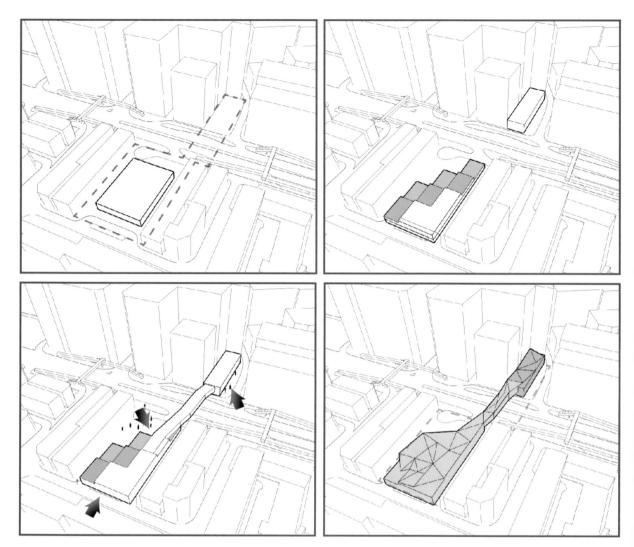

Sketch Up建模+Photoshop制作的形体变化分析图　设计者　朱启迪

（三）Photoshop 的水平要求

Photoshop 是设计师又爱又恨的一款软件，爱是因为它几乎无所不能，能够化腐朽为神奇，将毫不起眼的底图变成高大上的效果图。恨是因为要掌握 Photoshop 的强大技法需要长时间的练习和良好的个人美学素养。无论同学们基础如何，我们都建议将 Photoshop 作为作品集制图的支柱软件之一来对待。因为当我们有了优秀的方案，完善的技术图纸，接下来要锦上添花将方案惊艳地展示出来就离不开优秀的 Photoshop 制作。因此，我们建议同学们的 Photoshop 水平至少在建筑制图常用的技法层面应该达到优秀水平。先给大家看看 Photoshop 能做的哪些基本功能：

（1）强化光影。

（2）改变材质和纹理。

（3）增加设计细节。

（4）去除噪点和渲染错误。

（5）增加配景。

（6）调整场景色调和氛围。

（7）强调设计重点。

以下是两套图，前面是粗糙的渲染底图，后面是初步 Photoshop 加工后的修改初稿，大家可以看出来经过 Photoshop 加工后两张图从材质、色调和氛围都已经有了很大不同。

我们建议同学们在制作效果图的时候首先做好基础渲染，好的基础渲染底图有几个要素，这些是后期 Photoshop 所不容易修改提升的：

（1）良好的建筑黑白灰光影关系。

（2）重要的建筑和场景细节（精细建模）。

（3）整体的冷暖色彩基调。

（4）重要的材质表达以及贴图。

Photoshop作品之冷中带暖氛围　谢菲尔德大学　李怡萱

Photoshop作品之高冷氛围　米兰理工大学　李炫静

Photoshop作品之温暖氛围　墨尔本大学　刘宇恒

Photoshop作品之冷灰氛围　瑞典皇家理工学院　吴月

Photoshop作品之暖色氛围　瑞典皇家理工学院　吴月

Photoshop作品之高冷氛围　米兰理工大学　李炫静

（四）Rhino 的水平要求

对于喜欢曲面和有机形态设计的同学们来说，Rhino 以及相应的插件是非常好的伙伴。一些大体量的建筑，比如博物馆、体育馆常常可以用到 Rhino 建模和渲染。相对 SU 和 Photoshop 来说，作品集准备对 Rhino 及相应的插件（如 Grasshopper）要求并不那么高。因为大多数的设计还是走理性路线，我们不建议整本作品集都充满了科幻，参数化以及异形的设计。这样的路线是有非常大的风险的，除去少数院校的个别专业对参数化和异形先锋设计比较青睐，大多数国家的大多数院校还是很重视稳打稳扎的务实风格。宾夕法尼亚大学和伦敦大学学院的建筑系比较偏好参数化和先锋理念，但是在申请过程中递交的作品集可以完全没有这类风格的设计。扎哈与马岩松路线不适合所有人，而且也不容易掌握。如果同学们能够在有良好的设计理念、功能和造型结合紧密的前提下，适当地展示异形或者曲面的设计也是值得鼓励的（这类设计作品建议在作品集中不超过 2 个）。

说到 Rhino 就离不开小伙伴 Grasshopper 插件，Grasshopper 让 Rhino 软件的参数化可控设计变得十分灵活可控。因此在复杂建筑立面表皮，系统结构设计上面具有很大的优势。Rhino 还有很多其他小伙伴用于结构形变分析（Karamba），结构受力分析模拟（Robot），采光分析（Diva）等，这里就不一一举例了。这些表现和分析手法是增光添彩的，但不是必需的，使用的前提是同学们确实能理解这些软件，并且展示结果能够支持设计方案。

（五）Adobe Illustrator 的水平要求

Adobe Illustrator 是设计师画分析图的好帮手，Adobe Illustrator 也是 Adobe 绘图软件家族中强大的一员。但我们用到的功能只是很少一部分，也很好学，因此我们给出的建议是 Adobe Illustrator 达到熟练使用常用功能即可，可以足够配合 CAD、Photoshop 等其他软件。

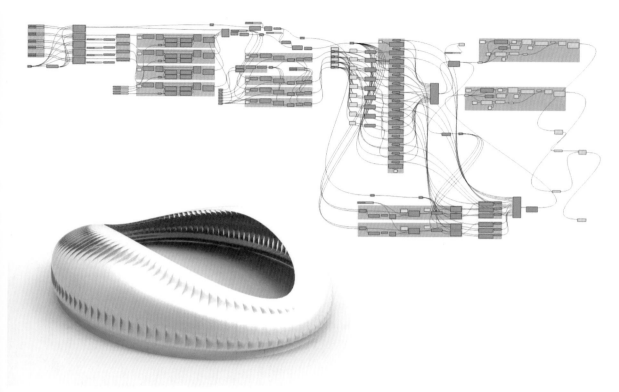

Rhino+Grasshopper有机形态建模　哈尔滨工业大学　鲍程远

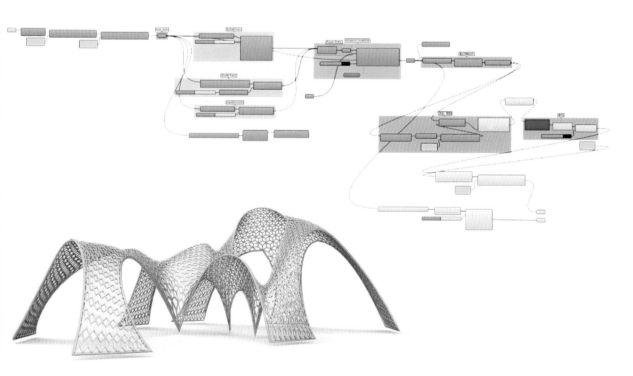

Rhino+Grasshopper有机形态建模　哈尔滨工业大学　鲍程远

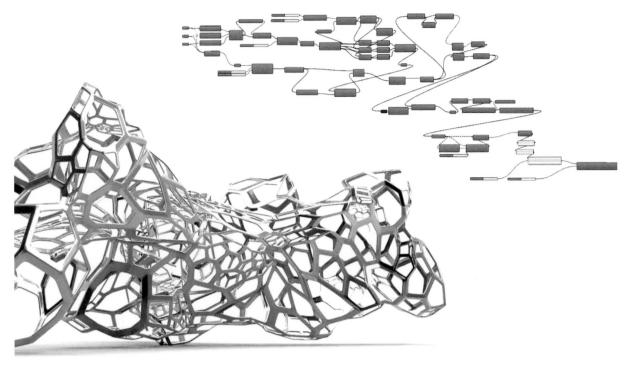

Rhino+Grasshopper有机形态建模　哈尔滨工业大学　鲍程远

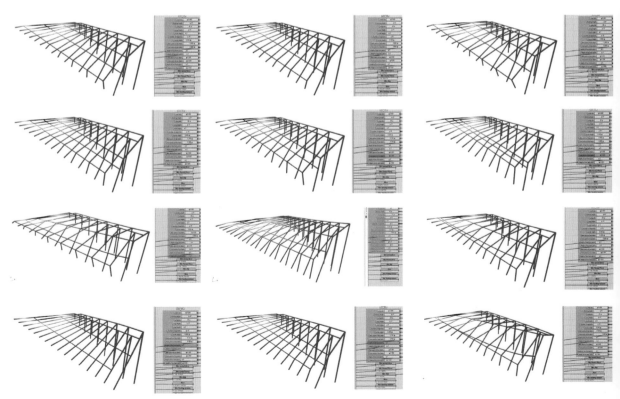

Rhino+Karamba结构形变分析　巅峰建筑学社

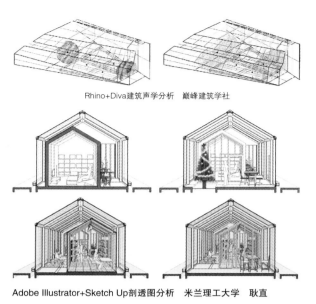

Rhino+Diva建筑声学分析　巅峰建筑学社

Adobe Illustrator+Sketch Up剖透图分析　米兰理工大学　耿直

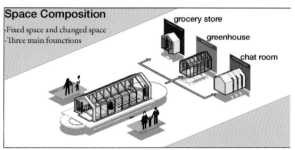

Adobe Illustrator理念分析图　米兰理工大学　耿直

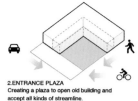
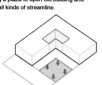

Adobe Illustrator理念分析图　米兰理工大学　许鹏辉

（六）其他软件的水平要求

除去我们介绍的这几类主要作品集支持软件，还有很多其他常用的软件。比如 3D Max、Revit 等。每一种软件都有它的优势和局限，作品集软件的采用可以不拘一格，根据申请者的自身喜好和方案的特征来选择使用。不管黑猫白猫，能抓老鼠的就是好猫。软件使用也是一样，十八般兵器难以样样精通，熟练掌握其中几种然后相辅相成就能达到非常好的效果。还是那句话，软件是表达思维的工具，如何决定软件使用和表达的还是我们的设计思维。

3D Max室内渲染　获得林肯大学室内设计offer　王彬琪

二、模型制作

模型制作是设计师的必备技能之一,模型分为很多种,作品集中我们常用的是概念展示模型、成果展示模型和节点(局部空间)模型。好的模型可以为作品集增色不少,各大事务所、艺术展都常常通过模型来展示自己公司的设计方案和理念。

丹麦伍重中心模型展

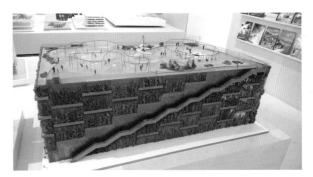

建筑模型 丹麦伍重中心模型展

景观模型 丹麦伍重中心模型展

景观模型 丹麦伍重中心模型展

(一)模型的材质选择

模型的材质选择对模型的表达至关重要,常用的模型材料有卡纸板、吹塑板、胶合木板、实木、金属、塑料(3D打印)、混凝土、黏土、石膏等。一般用于展示概念的模型常用卡纸板、吹塑版、黏土或石膏制成。成果展示模型多用纸板、胶合木板(激光切割)、塑料、实木等精细制成。

(二)模型制作途径的选择

模型制作途径无外乎手工切割、激光切割、3D打印、水泥或石膏浇筑、机械臂制作等等。

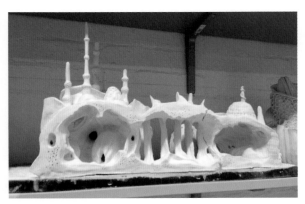

伦敦大学学院建筑系学生的石膏模型

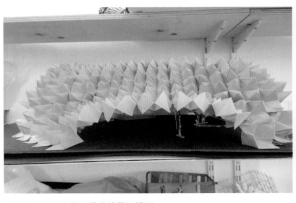

伦敦大学学院建筑系学生的草纸模型

诺曼福斯特事务所木质成果模型

英国皇家建筑师协会模型展　胶合木与实木

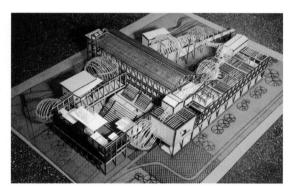

作品集方案展示模型　木材+塑料　激光切割　瑞典皇家理工学院
吴月

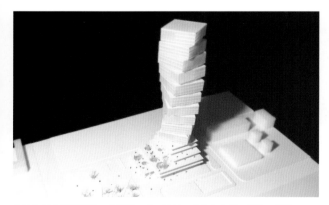

作品集理念展示模型　吹塑板　手工切割　瑞典皇家理工学院　吴月

作品集方案展示模型　胶合木板　激光切割　墨尔本大学　刘宇恒

三、手绘表达

手绘表达永远是建筑师的贴身技能，这是一项永不过时且全球通用的技能。好的手绘在方案生成阶段，理念沟通中有着至关重要的作用。境外教授都喜欢手绘好的学生，特别是现在计算机制图越来越占主导地位，手绘所体现的设计基本功，工匠精神，灵活创意等都一直是备受推崇的。因此我们在作品集里也可以多放一些精彩的手绘表达图作为辅助。

（一）概念草图

手绘表达在作品集中最常见的就是作为概念草图出现，这是最能体现设计者手绘基本功以及理念表达能力的图纸。添加得当会非常增光添彩。

（二）独立作品

另一种手绘表达是作为独立作品出现，通常放在正式作品后面作为附加页，展示设计者的手绘才能和美术功底。申请学院派风格以及对艺术修养重视的院校的作品集中较为常见。

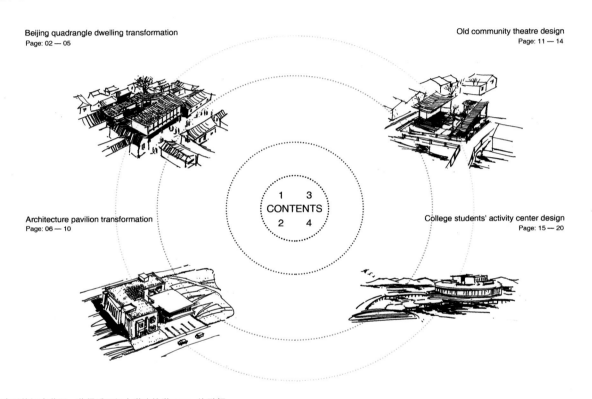

方案形体概念草图　获得爱丁堡大学建筑学offer　许鹏辉

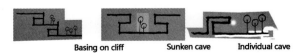

方案构思概念草图　获得伦敦大学学院建筑学offer　吴月

DRAWINGS

钢笔速写手绘表达　米兰理工大学　蔡笑革

OTHER WORKS

PAINTING BY MARKER

PAINTING BY PEN

彩铅、马克笔，钢笔表现　米兰理工大学　宋立群

水彩表现　　作者向畅颖

钢笔墨线水彩表现　　作者向畅颖

建筑水彩表现　　作者金梦潇

建筑水彩表现　　作者金梦潇

（三）辅助性分析图

另一类手绘表达是作为辅助出现在作品集当中的分析图，它们的角色是代替部分的 Adobe Illustrator 或者 Sketch Up 生成的分析图。包括形体生成、剖面分析、甚至节点大样的分析图，这类分析图往往更有叙述性和趣味性，通过手绘的形式让作品集图面更加生动。

CONCEPTS

1
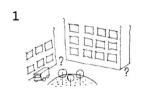
1.Use the corners and other leisure space inside the community

2
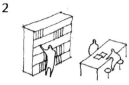
2.Reading can not only enrich people's life but also help them find friends with similar interest.

3

3.Cafe corner can attract more people for leisure and communication.

4

4.Mobile library which can easily been assembled and move are ideal for promoting communication.

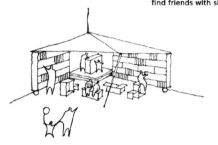

The small departments in gated residential communities like cages limiting the communication of people. In order to change the situation, we need to create attractive space offering chances for people to communicate and supporting activities preferred by various groups.

There are many unused leisure space between buildings in the community. If we could setup some community mobile libraries in these places, where people can easily access, the communication between neighbour can be improve, people can borrow and read books here, can have chatting and discussion here and also form new friendship here. Old books can be sent here and borrow to others, so resource can be also reused. The mobile libraries can move from community to community and there by building connections between communities.

手绘概念叙述图　墨尔本大学　吴越

第三节 留学作品集的内容准备

一、什么样的内容适合留学作品集

内容才是作品集的核心。什么样的内容最适合留学作品集呢？留学作品集以学术风格为主，所以最合适的是申请者深度参与过的（最好是独立制作的，这样不会有和其他申请者共用作品的情况，避免申请中很多院校可能产生的误会和麻烦），有特色的课程设计作业以及学术类的竞赛作品。如果有较好的实习和工作经历，也可以适当地放置一个左右的实际工程作品到作品集里。实际作品同样需要有详细的方案过程表达，理念分析以及避免过于商业化的效果图和分析图表达。

二、作品集任务书及内容筛选

（一）作品类型

作品的类型一般需要涵盖大学本科的主要知识点，以建筑学为例，全套作品需要从不同的类型来涵盖建筑空间、材料、结构、建筑与场地，建筑与城市文脉，建筑与生态环境等多方面的知识点。通过一套作品集向申请委员会展示申请者完备的知识结构体系。因此常见的作品类型可以包括别墅、幼儿园、小学、博物馆、美术馆、图书馆、高层办公、集合住宅、体育馆、城市设计、公园设计等。一般4个作品既有大型又有中小型设计。大型设计侧重体现对复杂体系和空间的把控，小型设计则可以

2016北京白塔寺建筑设计竞赛作品　　米兰理工大学　　李炫静

体现空间细节和申请者对构造和材质的理解和使用经验。设计作品跨越建筑、规划、景观的传统界限也是很好的选择，许多同学在准备作品集的时候往往还没有最终定下来自己想学的专业，那么一份作品集往往要投递一个或多个学校的相近专业。比如建筑的作品集也会用于景观建筑学和城市规划的专业申请。那么在作品集中有一两个设计体现多个专业的结合是很可取的做法。也会让评委老师对申请者的综合知识结构有更多的了解。

（二）作品规模

作品的规模和作品的类型是相关的。以建筑学为例，中小型的设计可以从几百平方米（别墅、幼儿园）到两三千平方米（小型美术馆、改造设计），大型设计一般也不要超过两万平方米。境外对大尺度的建筑设计侧重较少，较适中的设计规模也可以

借鉴传统建筑的设计　窑洞博物馆设计　获得谢菲尔德offer　吴月

借鉴传统土楼的设计　防灾幼儿园设计　伦敦大学学院建筑学　崔巍文

让作品集更具有深度和细节。

（三）作品的地域性和民族性

好的作品集除了体现作者自身的设计思想，设计风格以及完善的知识结构之外，如果能够对设计的地域性和民族性有所体现也是很受欢迎的。中国的学生面对的是来自全世界各地的竞争者。设计的风格和背景千差万别，如何让自己的作品从众多的竞争者中脱颖而出，地域性和民族性的特色会提供很大的帮助。中国传统建筑和规划传承了数千年古老的中华文明，有许许多多值得我们去好好研究、发掘和发扬光大的地方。从建筑构造、建筑材料、群体建筑布局、意境空间营造、建筑与环境融合等许多方面都备受境外设计大师称道。因此，同学们的设计可以一定程度地体现对自己国家和民族文化的思考和再提炼。

三、结合院校以及自身背景制定作品集框架

最后，作为这一章的结尾。我们再次强调结合院校要求和自身背景制作作品集的重要性。好的设计一定是发自内心的，结合自己的设计特点和喜好才能做出有真情实感的设计，首先要打动自己，才能打动别人。其次，作品集的材料也需要认真根据目标院校的要求来准备，很多院校会将作品集的详细要求（甚至采分点）公布在申请的官网上。因此，在准备之前详细阅读院校的要求才能做到心中有数，有的放矢。

以代尔夫特理工大学建筑学申请为例，学校详细列举了作品集的要和注意事项（https：//www.tudelft.nl/en/education/programmes/masters/architecture-urbanism-and-building-sciences/msc-architecture-urbanism-and-building-sciences/admission-and-application/non-dutch-bsc-degree/）。

作品集整体要求：

（1）至多不超过5个作品。
（2）不超过两个实际工程作品。
（3）至少一个完全独立的设计作品。
（4）包含毕业设计或最新的设计。
（5）至少一个设计能清晰的体现申请者的建筑

What to include in your portfolio

The portfolio cover should include your full name and contact information.

Number and nature of projects:

- Include a total maximum of 5 projects.
- Include no more than 2 professional projects.
- Include at least one individual project.
- Include your final project (if you have not yet completed it, include your most recent project).
- At least one of the projects included should clearly demonstrate your capabilities with respects to structural and technical project viability. We highly recommend that these structural and technical aspects are exhibited in either the final thesis project, or alternatively, your most recent or most advanced level, project.
- Include a summary of your motivation letter of no more than 400 words (make sure all elements of the essay are covered in this summary).

结构和技术知识水平。建议在毕业设计或者最近的设计中体现出来。

（6）包含一篇400字以内的动机信。

每个作品的详细要求：

（1）包括概念草图，理念的演化以及设计者如何整合设计思路的过程。

（2）详细的最终展示图纸，通过条理清晰方式展示出来。

（3）相关的理念文字介绍，包括设计所受的启发，最终的设计目标和意图。文字需要精练。

 Project documentation

Per project include:

- concept sketches, showing preliminary visual (and verbal when relevant) development of ideas, approaches and methods and showing how you organise your ideas.
- finished drawings of original design work, documenting in a clear and precise manner both the intention and the resolution of each project.
- a written explanation (in English) of the design concept and solution Here you should include the inspiration behind the project as well as the goals and objectives of the design. Be clear and concise. Longer texts should be worked out in the form of your essay.

Please note that for all drawings we require you to include the name of the author.

第三章
留学作品集要素详解
Detailed Introduction Of Portfolio Elements

作品集方案要素
Design Concepts Elements

第一节

一、平面图

平面图（plan /plan layout）是建筑/规划类专业作品集中的重要组成部分。是最基础也是最重要的图纸之一。好的平面图需要做到：①合理的功能布置；②清晰的流线；③重点空间的强调表达；④明确的室内外空间关系；⑤正确及简洁的技术表达。

平面图与剖面图（包括剖透图）都属于让阅读者理解设计方案内涵的核心图纸。当初步的方案展示让阅读者眼前一亮后，接下来就是期待在平面图找到其心理期待的相应空间展示。抑或当读者对方案产生困惑之后，第一反应往往也是仔细查阅平面图来期待找到解答疑惑的答案。好的平面图也能让审核者第一时间看出申请者的设计功底，如果说效果图、分析图很多时候可以将普通的设计"升级"成"高大上"的设计，那么平面图很多时候就会把普通的设计或者有缺陷的设计直接打回原形。好的平面图耐看，有层次，有策略，朴实而能挑起大梁。不好的平面图常常使人产生混乱、花哨、漏洞百出、捉襟见肘的感觉。

具体需要怎样表达作品集里的平面图呢？作品集由于篇幅、版面的限制，对平面图的处理也和平时作业图纸的平面图有很大区别。第一点是要去粗取精，有别于作业图纸和初次设计，施工图的详细平面图。作品集的平面图往往都非常精炼，可以不标注尺寸、轴线、简化门窗的表达。只需要在重点空间和想要详细表达的区域布置室内家具等。

简化的平面图表达适用于大型和中型体量的设计，如高层建筑的标准层、医院、学校、博物馆等方案。当设计有许多平面但并不需要每层都详细表达时即可用简化的方式来处理平面。但是很多时候我们也希望并且需要有更详细的平面表达，在作品集里用较多的篇幅来浓墨重彩地表达丰富和有特色的设计空间。这时候我们就需要用强化重点的平面图来代替简化的平面图。重要的建筑结构、室内外空间变化、家具布置、景观设计甚至建筑材质都可以强化表达，门窗、轴线和尺寸线依旧可以简化。

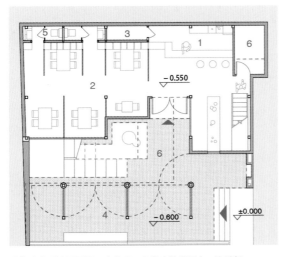 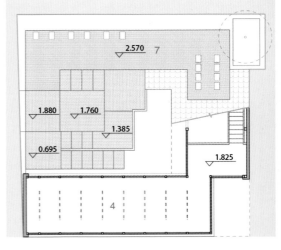

院落改造设计平面图　米兰理工大学建筑学硕士　许鹏辉

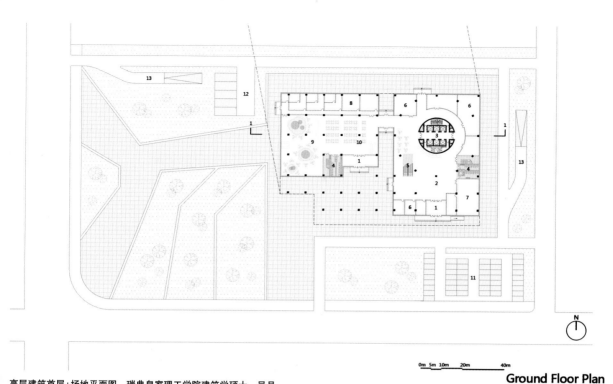

Ground Floor Plan

高层建筑首层+场地平面图　瑞典皇家理工学院建筑学硕士　吴月

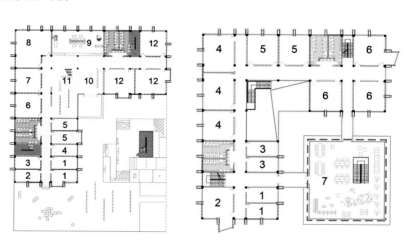

建筑系馆平面图　米兰理工大学建筑
设计专业硕士　许鹏辉

PLAN

Fifth Floor Plan

1　Elevator room
2　Leisure space
3　Hotel room

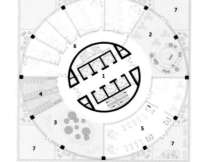

Eleventh Floor Plan

1　Elevator room
2　Cafe
3　Leisure space
4　W.C.
5　Gym
6　SPA center
7　Exterior leisure space

高层塔楼标准层平面图　瑞典皇家理工学院建筑学硕士　吴月

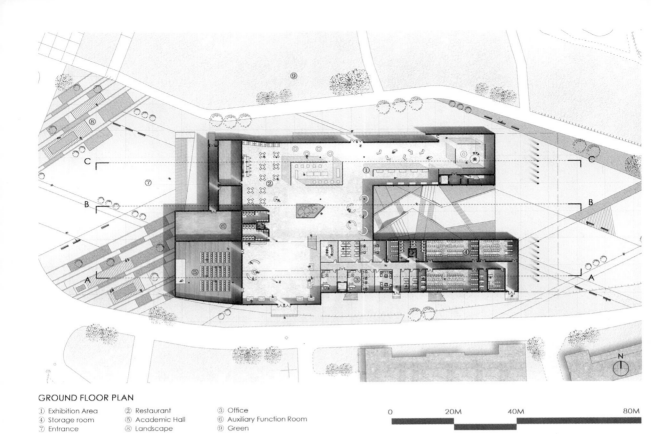

GROUND FLOOR PLAN
① Exhibition Area　② Restaurant　③ Office
④ Storage room　⑤ Academic Hall　⑥ Auxiliary Function Room
⑦ Entrance　⑧ Landscape　⑨ Green

博物馆首层+场地平面图　米兰理工大学建筑学硕士　李炫静

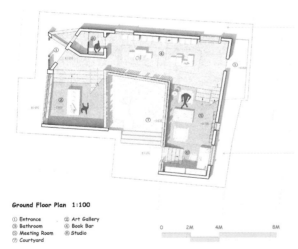

Ground Floor Plan 1:100
① Entrance　④ Art Gallery
② Bathroom　⑤ Book Bar
③ Meeting Room　⑥ Studio
⑦ Courtyard

小住宅首层平面图　米兰理工大学建筑学硕士　李炫静

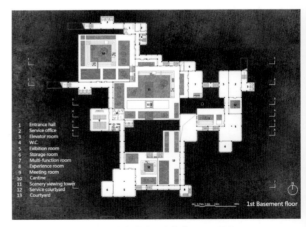

1 Entrance hall
2 Service office
3 Elevator room
4 W.C.
5 Exibition room
6 Storage room
7 Multi-function room
8 Experience room
9 Meeting room
10 Cantine
11 Scenery viewing tower
12 Service courtyard
13 Courtyard

1st Basement floor

博物馆平面图　获得伦敦大学学院　建筑学offer　吴月

二、立面图

　　立面图（elevation）是传统的建筑图纸。主要的作用是展示建筑各个朝向的形式、材质和建筑语言。在建筑作品集中重要性略低于平面图，主要是由于在表现手法多样化的今天，立面图作为体现建筑形态的作用已经逐渐被多样化的人视角透视图、鸟瞰图、模型效果图所取代。因此同学们在作品集中不用浓墨重彩地表现立面图（前提是有很多其他图纸能够清晰地展现方案的三维效果）。不过这并不意味着立面图就不重要，只要是放到作品集中的图，都需要认真表达。建筑立面图除了能让人直观感受设计者对材料的选择、建筑比例的控制之外，还能清晰地反映出建筑尺度与周边环境的关系。特别是在高密度区域的建筑设计以及原有老的街区立面图,对体现设计与周边环境的关系、风格材质的延续或对比、街道与公共开放空间的尺度变化均非常重要。

三、剖面图

剖面图是作品集中最重要的一类基本图纸。剖面图又细分为普通剖面图和剖透图。剖透图根据透视角度又可以分为人视点、鸟瞰和仰视剖透图(路易斯·康最喜欢的一类剖透图),根据透视种类可以分为轴测剖透图和三维透视剖透图。剖面图之所以重要在于它可以非常直观地把建筑内部空间变化、体量虚实关系以及结构体系展示出来。很多时候设计会始于剖面,优秀的设计师可以直接从剖面入手开始设计。好的剖面体现的是设计者对空间的理解和表达,对三维空间的想象,对结构的详细考虑和选择。因此剖面图在作品集里占有非常重要的地位,复杂的建筑往往需要多个剖面图来表达。建筑整体剖面图、局部剖面图、建筑加场地剖面图分别能展示建筑的整体内部空间关系,局部空间细节和建筑内部空间与周边城市或自然环境的延续性。接下来就让我们看一些优秀的作品集剖面案例。

简洁的立面图体现了材质和建筑体量虚实的变化　米兰理工建筑设计专业　耿直

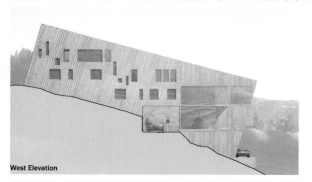 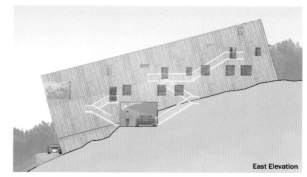

建筑立面图　体现建筑与周边环境的关系　新南威尔士大学建筑学　尹媛　　　建筑立面图体现建筑与周边环境的关系　新南威尔士大学建筑学　尹媛

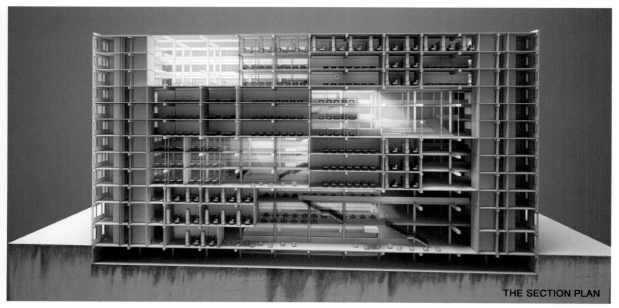

教学楼水平视角剖透图　墨尔本大学　朱冠宇设计作品

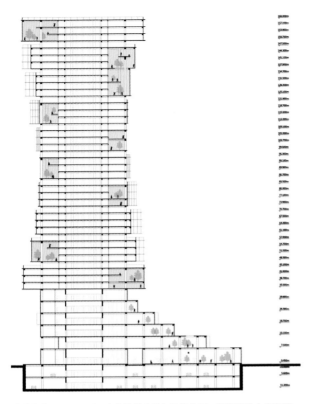

简洁的高层剖面图展示建筑体量变化与结构体系　获得伦敦大学学院建筑学offer　吴月

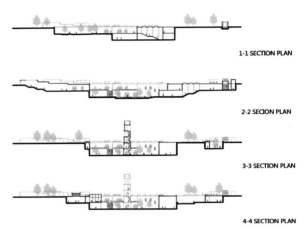

组合剖面图展示博物馆建筑实体与院落以及周边环境的关系　获得伦敦大学学院建筑学offer　吴月

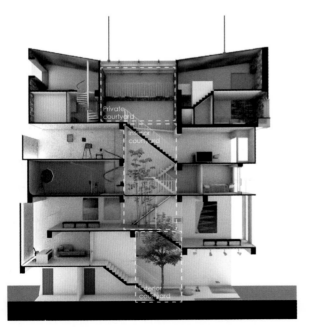

水平视角的剖透图展示复合功能建筑的组合嵌套关系　香港中文大学建筑学硕士　朱启迪

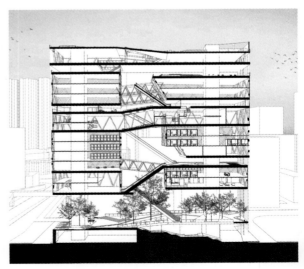

水平视角的线稿剖透图展示建筑的基本结构和空间转化　香港中文大学建筑学硕士　朱启迪

水平视角的剖透图　展示建筑室内丰富的空间以及建筑与场地的关系　米兰理工大学建筑学硕士　李炫静

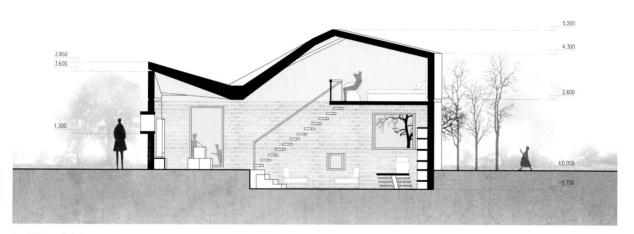

细致的小尺度建筑剖面图体现建筑的结构形态和具体空间尺度　米兰理工大学建筑学硕士　李炫静

四、总图

总图（master plan）是许多同学容易忽视的图纸，无论是建筑、规划、景观专业，总图都是介绍方案理念的基本要素。每一个设计都有它的独特性，独特性产生的很重要一个因素就是环境对设计的影响，包括不利制约和有利条件。无论是风景优美的自然环境还是高密度的城市 CBD 中心，场地都会对设计带来极大的影响。而境外的建筑院校非常重视设计产生的逻辑性，欧洲的院校非常看重建筑和景观与周边已有城市环境的关系，建筑的层高体量、形态、材质选择，入口以及场地设计都与周边环境息息相关。北美的院校更希望看到设计者对方案的生成有详细和清晰逻辑性的展示。因此，总图作为呈现设计与现有周边影响制约因素关系的首要图纸，就显得十分重要。下面给大家展示不同专业的优秀总图。

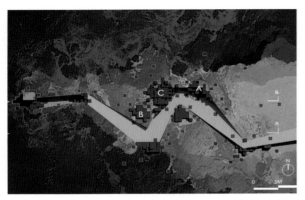

纪念碑建筑设计方案总图 展示建筑体量与路径，人工设计与自然环境之间的关系　米兰理工大学建筑学硕士　李炫静

建筑改造项目总图 新设计与旧城市肌理之间的关系　米兰理工建筑设计　耿直

建筑改造项目总图 展示设计场地与周边自然环境，已有基础设施之间的关系　瑞典皇家理工学院　建筑学硕士　吴月

滨水城市更新设计总平面图 体现景观体系，公共空间体系以及建筑肌理之间的关系　瑞典皇家理工学院　吴月

博物馆总平面图 展示新建筑与周边已有建筑组群，建筑与环境，建筑场地与周边道路关系　米兰理工大学建筑学硕士　李炫静

五、分析图

分析图可以说是作品集中最精彩的部分。分析图种类繁多，以建筑方案分析图为例，主要包括：

（1）方案理念分析图。

（2）建筑分析图。

（3）技术分析图。

方案理念分析图往往包括场地分析图、概念演化分析图、形体生成分析图等。建筑分析图则涵盖功能分析图、流线分析图等。技术分析图包括材料分析图、结构分析图、节点大样图、生态分析图等。分析图往往在作品集中占约三分之一的内容比重，和平面图、剖面图、效果图、文字相互结合，起到相得益彰的效果。

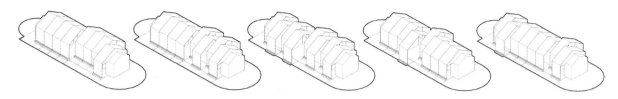

理念分析图之建筑形体演变　米兰理工大学建筑设计专业　耿直

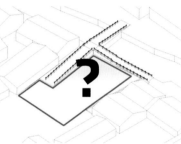

1.SITE
Base is a piece of open space near the water's edge and Bridges.

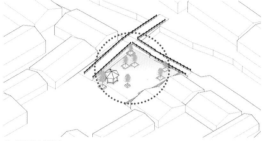

2.REMAIN
Base retain the ancient bridge and pavilion.

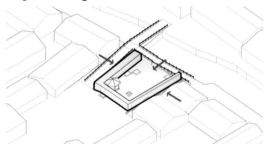

3.VOLUME
Bridges and houses to surround close to determine the scope of the theatre.

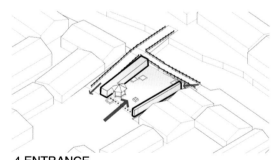

4.ENTRANCE
According to the site traffic sources to determine the theatre main entrance.

5.HEIGHT DIFFERENCE
According to the terrain ups and downs build variety show and activity space.

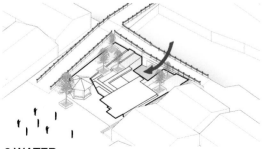

6.WATER
The introduction of water enriched theatre in landscape and irrigation of the plants.

理念分析图之建筑形体生成　米兰理工大学建筑设计　许鹏辉

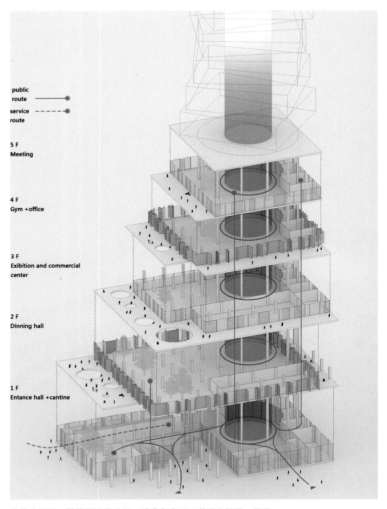

建筑分析图之功能分析图　米兰理工大学建筑设计及历史专业　宋立群

建筑分析图　爆炸图流线分析　瑞典皇家理工学院建筑学　吴月

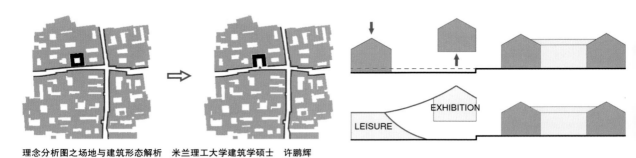

理念分析图之场地与建筑形态解析　米兰理工大学建筑学硕士　许鹏辉

　　分析图在作品集中有着举足轻重的分量。同学们一定要根据自己的方案特点来安排分析图的数量与种类，并且与其他图纸相互配合，思路清晰、逻辑顺畅地表达自己的方案设计，将掌握的设计知识从不同角度展示出来。

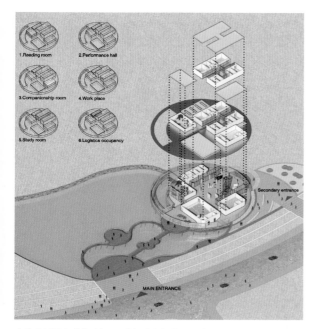

建筑分析图之功能分析图+空间分析爆炸图　米兰理工大学建筑学硕士　许鹏辉

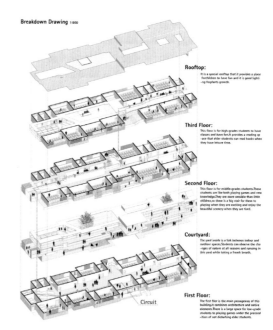

建筑分析图之功能爆炸图　谢菲尔德大学建筑学硕士　李怡萱

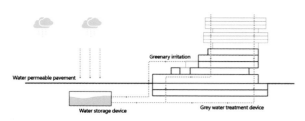

技术分析图　可持续水循环设计示意图　瑞典皇家理工学院建筑学　吴月

技术分析图　共享空间日照与通风分析图　瑞典皇家理工学院建筑学　吴月

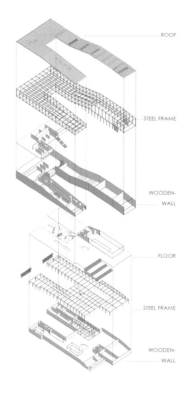

技术分析图之结构体系爆炸图　米兰理工大学建筑学硕士　李炫静

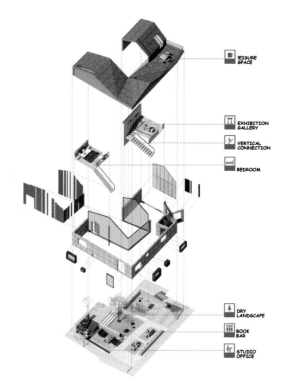

六、效果图

效果图无疑是作品集中所有图纸中最引人注目的图纸，也是许多同学深深为之困扰的图纸。怎样的效果图才算好的效果图呢？角度恰当，能展示设计最佳姿态，色调稳重而朴实是好的效果图的基本要素。效果图的风格可以有很多种，模型风格的，写实风格的，拼贴风格的都值得推崇。要避免效果图过于商业化，夸张绚丽华而不实。学术感比较强的效果图是境外院校普遍比较欣赏的效果图类型。

写实风格的清新效果图　米兰理工大学建筑学硕士　李炫静

拼贴剪报风格的效果图　香港中文大学建筑学硕士　朱启迪

模型效果的学术效果图　墨尔本大学建筑学硕士　刘宇恒

❤ 空中娱乐场
Casino in the sky

❤❤ 铁道游览车
Railway observation car

❤❤ 电线猫空缆车 ❤
Cable car of electrice wire

❤❤❤ 山墙悄悄话
Whisper on the wall

❤ 地下音乐亭
Music pavilion of underground

❤ 屋顶共享花园 ❤❤❤
Shared roof garden

改造成果展示

漫画风格的清新效果图　作者郑权一

水彩风格的清新效果图　米兰理工大学建筑学硕士　许鹏辉

模型风格的效果图　米兰理工大学建筑学硕士　许鹏辉

写实风格的室内效果图　获得威斯敏斯特大学室内设计offer　王彬琪

写实风格的学术效果图　米兰理工大学建筑学硕士　李炫静

写实风格的室内效果图　获得林肯大学室内设计offer　王彬琪

写实风格的景观效果图　米兰理工大学可持续建筑与景观硕士　李同学

七、其他

除了主要的图纸之外，作品集的文字、封面、目录及扉页都是需要仔细斟酌的。文字是作品集中非常重要但是特别容易被同学们忽视的一部分。文字承载着图片所不能代替的功能。适度的文字解说可以让设计更清楚地展示给阅读者，将图面所包含的含蓄理念更直截了当地再次阐述。同时也增加作品集阅读的趣味性。同时文字本身也是一种版面的构成，字体选择、标题大小、正文大小、文字排版都很有讲究。具体选择什么字体可以因人而异，但必须清晰易懂，不能过于花哨，也不能过于呆板。

既不会喧宾夺主抢去图纸的风头也不会太过于平淡让人觉得索然无味。字体大小和作品集版面大小，文字数量相关。以能看清楚而不夸张为宜。以A4纸张大小为例，正文字体可以选择9到12号字体。过小不易看清，过大显得呆笨。

目录是让读者理解作品集结构的开篇介绍，需要简洁明了。但在尽量简洁直接的同时也要避免过于单调。

在篇幅允许的情况下，给每个设计加上单独的扉页是一件很棒的事，可以清楚地将不同设计分隔开。同时可以对接下来的设计做一个简单的介绍。

所以除去最主要的图纸、文字、扉页、目录都不可轻视。同学们最好多下一番功夫，将作品集做到精益求精。

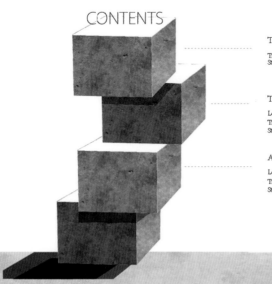

Architecture designing is not the privilege of architect. We design an architecture which is divided into two parts. One is the entity space providing places for daily education activities and the other is the frame space which provides a possibility for students to design by twhemselves.

The Unfinished Project
PROJECT INFORMATION:
Time : september.2015-november.2015
academic work

DRAWINGS:
Pespective plan: (Sketch up, VRay and Photoshop)
Analysis plan: (Adobe Illustrator,Sketchup and Autodesk CAD)
Diagrammatic figure: (3mm Cardboard and Photoshop)
Master plan: (Autodesk CAD and Phtotshop)
Profile map: (Sketch up, Autodesk CAD and Adobe Illustrator)

作品集的表达
Expression Of Portfolio

第二节

说过作品集的主要要素之后就是如何将这些丰富的内容编织成引人入胜的故事了。接下来我们将介绍一些常用的表达技巧和实用方法。

一、单个作品表达

每一个作品都是一个完整的故事。在独立作品表达的时候，我们需要站在阅读者的角度来思考，既要通过充分的信息阐述让读者熟悉设计背景、立意、新颖点和特色，又要考虑到从阅读者的专业角度来避免过多冗余的信息展示。就像文学作品一样，作品的表达可以开门见山直截了当地先展示作品的特色和重点，然后对设计缘由、方案理念娓娓道来。也可以先做充分铺垫和陈述，按照设计发展的顺序来展示方案从理念一步步变成具体的设计。这两种都有各自的优势，同学们可以按照自己的风格和爱好来采用。第一种往往给人更清晰的印象和冲击力，第二种更适合着重想表达对设计的认识和思考的同学。无论哪一种，都需要有完整的体系，起承转合，抑扬顿挫。图纸的顺序以及配合方式都要始终服务于突出自己设计重点特色，展示贯穿整个设计的主题概念，将设计者所掌握的知识最佳地表达出来。

二、全套作品集表达

单个作品就像小说的章节，全套作品应该有一个统一的风格和脉络。这里所说的风格及脉络可以包括连贯演变的设计思路、不同设计所包含的共同关注点等。整套作品集的风格除了和申请者自身特色相关之外，还和申请的具体专业有直接联系。了解具体申请专业的要求和特点有的放矢地安排作品是最合适的，比如申请建筑和城市设计结合的专业就需要考虑建筑设计作品和城市设计作品的结合性。有时候我们会建议两个作品是承接关系，比如前一个设计是街区改造的城市设计，后一个设计是前一个设计街区里的重点建筑设计。申请可持续建筑设计方向的作品集最好在不同的作品里体现对可持续发展的理解思考和适当的技术运用。申请建筑历史保护方向的作品应当从不同城市层级，从宏观到微观体现对历史文化的尊重与思辨，新旧的对立抑或统一。

（一）作品难度的递进性

作品集的作品难度排列可以采用递进式，同时也符合时间顺序将较简单难度较低的设计排在前面，然后逐渐递进。另一种是穿插型，在开篇可以放上较为复杂的设计，给审阅者比较深的第一印象，接下来排列较为简单的设计，最后再加上复杂的大设计。需要注意的是，不建议将最简单的设计放在最后两个设计的位置。均匀穿插搭配以及难度递进式的排列都要强于虎头蛇尾的排列方式。

（二）作品类型的搭配

作品集的作品搭配应当遵循全面而有重点的原则。全面，是在可能的情况下尽量能体现申请者全面的设计知识，从设计理论，风格喜好到对空间和材质以及最新技术的了解和认识。因此可以包含从低年级到高年级的不同种类设计，内容和篇幅上以高年级的复杂设计为主。以建筑设计作品集为例，可以包含别墅、幼儿园、博物馆、图书馆、高层建筑、体育馆等不同类型不同特色的建筑方案。在此基础上，申请者可以结合自己的设计特色和偏好来调整。比如对建筑与城市环境情有独钟的同学可以通过不同城市环境中的住宅、公共建筑、新建筑设计或旧建筑改造来体现自己对建筑与城市的理解。对生态可持续感兴趣的同学可以结合不同的作品，从不同建筑环境层级来展示相关的设计思考。比如从建筑材料的可持续性研究到城市设计尺度的绿地系统与城市公共空间的相互交融等。

第三节 作品集排版及配色

一、作品集排版

在作品集的排版上同学们可以八仙过海各显神通。除了严格按照申请院校的要求来准备之外就可以按照个人喜好随心所欲了。不过前提是一定要先了解申请院校对作品集格式的具体要求，不符合学校明文规定的作品集很可能不被受理，哪怕设计作品非常优秀。我们会详细介绍不同院校对作品集的详细要求，在这里就不过多陈述。很多同学会申请不同的院校，这种情况下可以先制作一个作品集蓝本，然后针对不同院校的要求来调整版式。下面就一些常用的排版方式给大家做一些建议。

横版（Landscape orientation）

横版是留学作品集最常用的排版版式，很多学校甚至直接规定作品集需要做成横版 A4 或 A3。学生时代课程作业交图往往都是竖版 A0，A1 甚至更大的连续展板。很多同学对突然缩小到 A4 版面的排版会不太适应。其实背后的原理很好理解，无论是留学申请还是求职申请，审核者往往都是在计算机屏幕上浏览作品集，这样 A4 大小的版式与计算机屏幕大小是最匹配的，可以保证在不用缩放的情况下舒适地浏览审查。横版的另一个好处是适合布置效果图，选择横版排版之后具体的布置就可以因人而异，遵循清晰简洁适合阅读的原则即可。

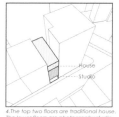

横版分析图排版　香港中文大学建筑学硕士　朱启迪

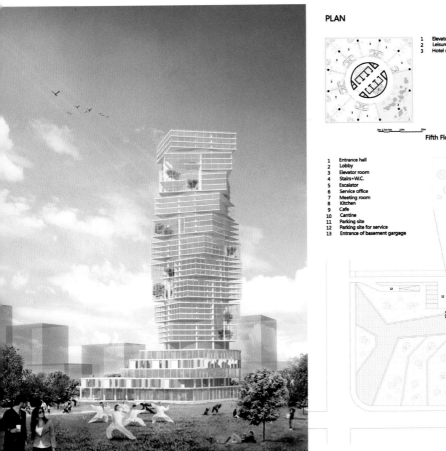

横版　效果图加平面展示　瑞典皇家理工学院建筑学硕士　吴月

竖版（Portrait orientation）

如果申请院校没有指定横版作品集排版，非常喜欢竖版排版的同学也可以用竖版（特别是如果需要邮寄打印的作品集的话）。不过由于之前提到的电子版作品集更适合横版排版便于在计算机上阅读，因此我们在此不太推荐竖版排版。

特殊排版

除去常见的A4纸张比例的横版或竖版排版，有的同学也喜欢用小一号的纸张大小页面来设置作品及尺寸，如B5大小或者4∶5比例的近似方形页面。方形的页面看上去更加新颖，适合个性化比较强的同学使用，缺点是每张所包含的内容由于版面大小和比例相对A4页面会有所减少。

二、作品集配色

作品集的配色可以选用暖色系或冷色系，以清新淡雅的风格为佳。明快的素雅的色调可以更好地突出设计的本质和表达的重点。高级灰颜色系列是常用的主打色彩，尽量不要过多使用过于花哨酷炫的鲜艳颜色。灰色和黑色作为局部背景色的使用可以让图面更加稳重。一套作品集不同的作品可以都用暖色调或者冷色调，也可以冷暖兼顾，但尽量不要使用对比过于强烈的色调。

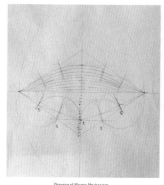

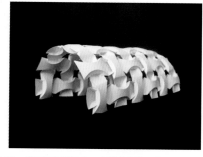

长宽比例4:5的页面比例　美国得克萨斯农工大学建筑学硕士　黎子阳

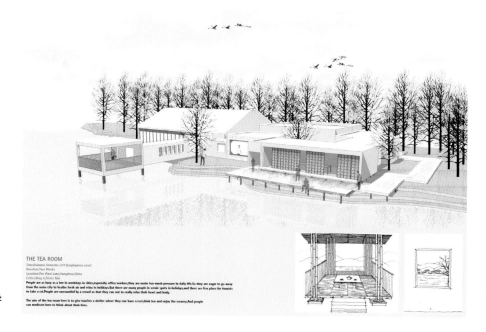

暖灰色调的图面　谢菲尔德大学
建筑学硕士　李怡萱

暖色调的图面表达　香港中文大学
建筑学硕士　朱启迪

69

冷灰色调图面　米兰理工建筑学硕士　李怡萱

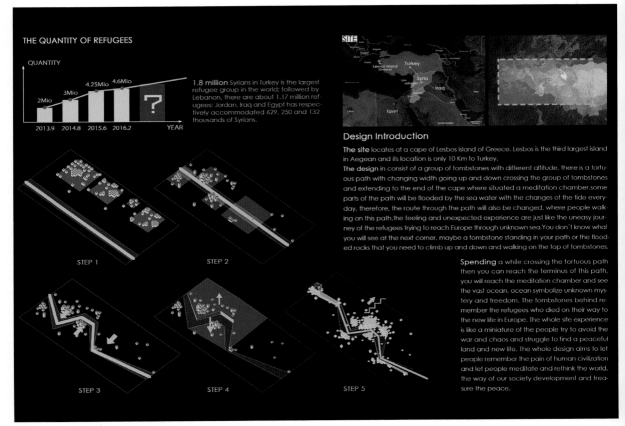

黑色背景色图面　米兰理工大学建筑学硕士　李炫静

第四章

优秀作品集案例及院校要求详解

Outstanding Portfolio Exemplars And Requirements By Different Universities

优秀作品集案例展示 Outstanding Architectural Portfolio Exemplars

北京白塔寺院落改造设计
（米兰理工大学）
米兰理工大学建筑学硕士李炫静
（白金奖获得者）

第一节

一、建筑类专业方案

该设计选取"北京小院的重生"2016白塔寺院落更新国际方案征集设计任务书。对北京白塔寺周边胡同区的一个小四合院进行重新设计，使其具备居住、工休闲、娱乐的复合功能，并在狭窄的场地内保留了传统的院落空间设计。在兼顾现代生活方式的同时，用传统的建筑语汇诠释了北京小院儿的特色与内涵。

Period: March.2016
Acedemic Year: Fourth
Location: Beijing, China
Site Area: 144 m²
Type: Competition | Individual Work

01 REINVENTING OF CO
-TRANQUIL TIME CAPSULE

院落鸟瞰与胡同肌理分析

别墅小住宅类设计作品
Villa/House category

Different from the typically ordered Beijing courtyards, there are many tiny courtyards scattered in the old city of Beijing.
"Yuan'er" (Courtyard), is the most essential element in the composition of the old town of Beijing. For a long time, "Yuan'er" contains all the possibilities of working, entertainment, leisure and dwelling for the local people. The space inside is usually very small, and living in it can be hard yet colorful. It is part of Beijing's realistic life, and is a miniature of China's old cities

别墅小住宅类设计作品
Villa/House category

VIEW OF ENTRANCE

With societal advancements, people have realized that standard large-scale demolition and construction of Beijing's urban core is no longer viable. In its stead, small-scale, organic renovation models are attracting greater attention and use.

新院落街景透视图

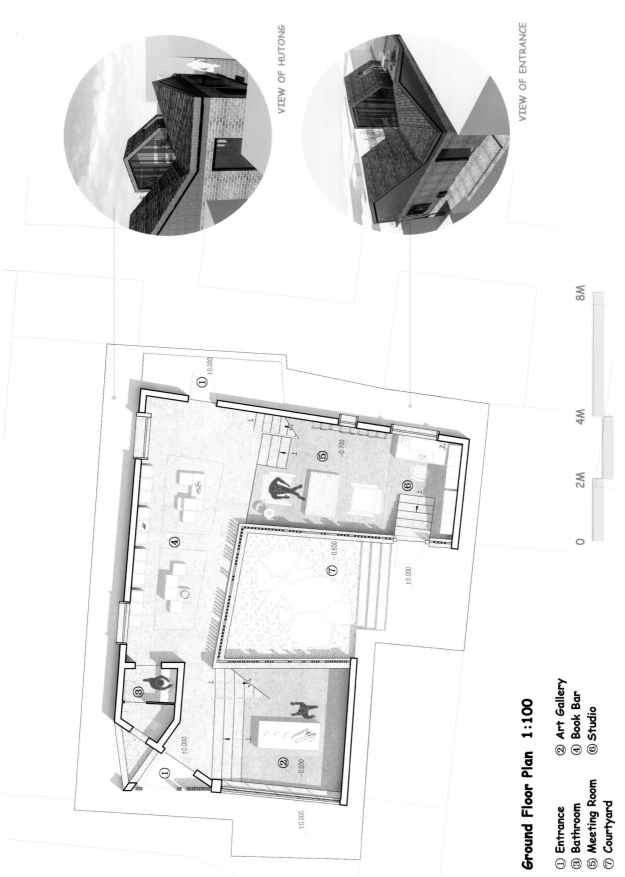

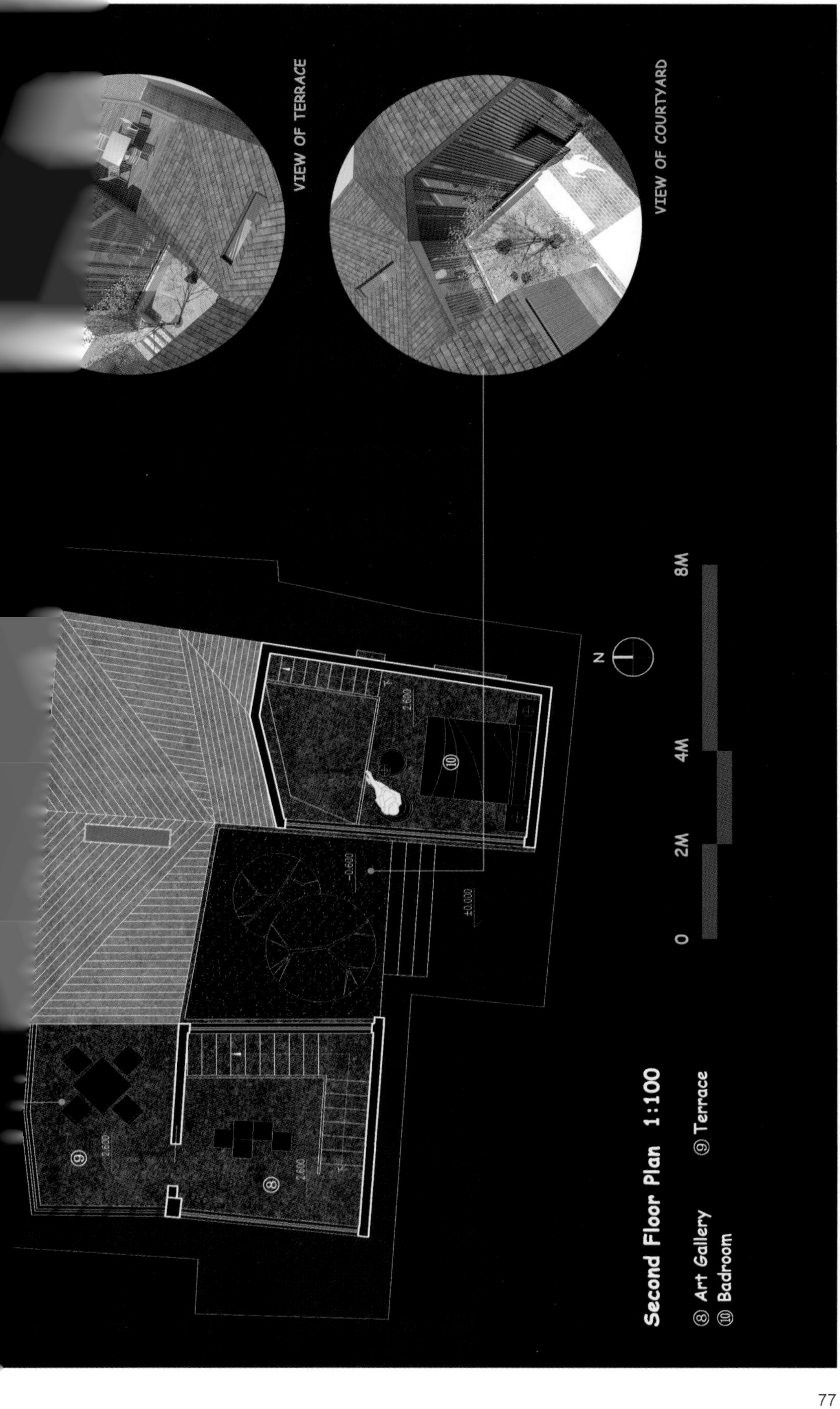

别墅小住宅类设计作品
Villa/House category

VIEW OF TERRACE

VIEW OF COURTYARD

Second Floor Plan 1:100

⑧ Art Gallery ⑨ Terrace
⑩ Badroom

二层平面图与院落营设计

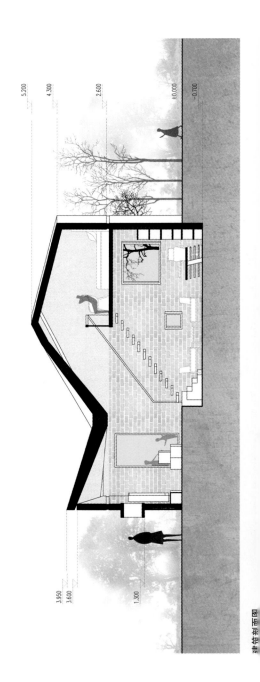

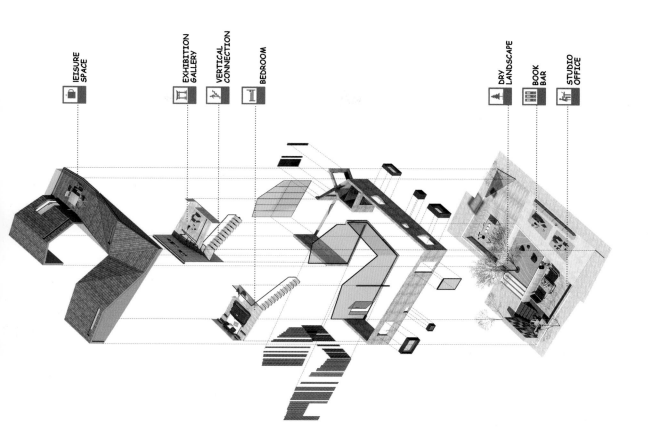

室内空间展示与建筑爆炸图

香港摄影师工作室兼住宅设计（香港中文大学）

香港中文大学建筑学硕士　朱启迪

本设计选址在中国香港 Yik Yam Street, Happy Valley，是位于两栋居民楼之间的狭长地块。本设计的挑战在于基地的狭小，采光环境的局限以及两侧建筑的压迫感。在有限的条件下要解决摄影师的工作室空间、居住空间、办公空间以及作品展和尽可能提供公共交流场所的可能性。

Finished time: November 2014
During the second year of undergraduate program of Architecture design
Individual work
Location: Yik Yam Street, Happy Valley, Hong Kong

街景效果图

-Modern studio & Traditional house

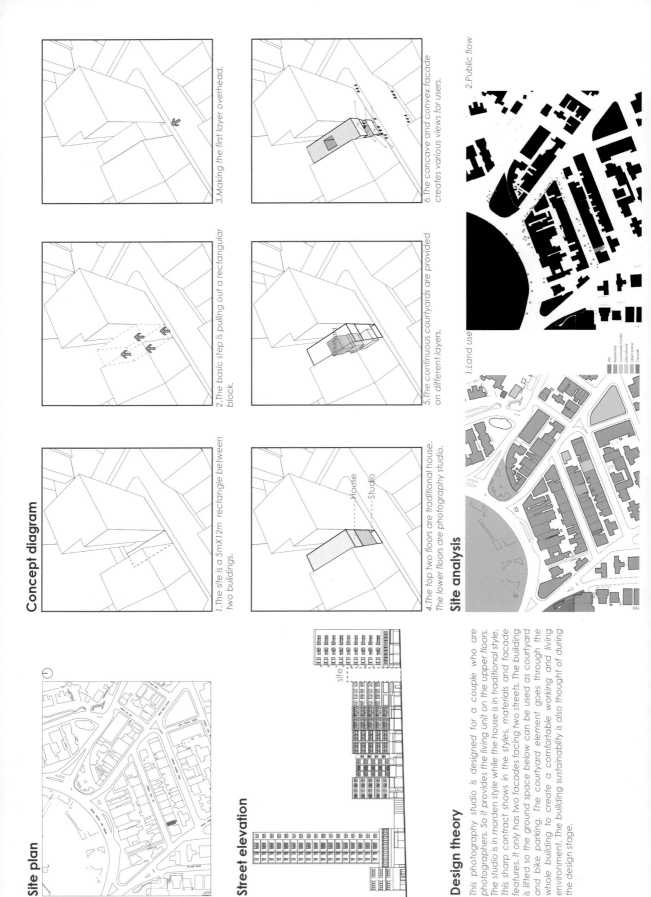

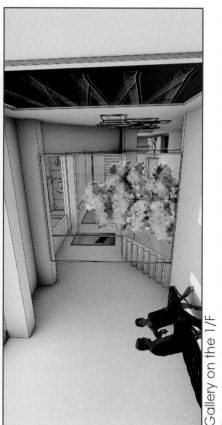

Entrance

The small courtyard is on the ground floor which provides the greenery for the studio, playing space for children from neighborhood residence and the bike parking.

Gallery on the 1/F

Photography works can be shown in the gallery, so the cleints would enjoy the fantistic arts and the middle courtyard while they are going upper to the workshop or wating area to take pictures.

Plans

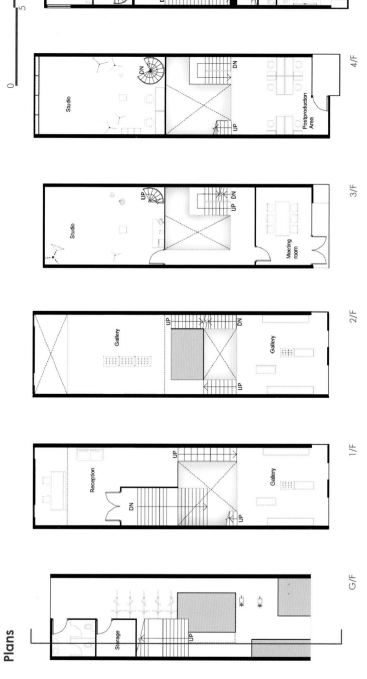

入口空间，展示空间以及各层平面图

Section

Green roof
This green roof not only retains rainwater, but also moderates the temperature of the water and act as natural filters when it runs off to the storm tank on the ground floor through pipes inside walls.

Flat roof skylight
The skylight directs to the interior space through flat roof with openable glass windows. The shading device can modify the sunlight and natural ventilation well.

Private courtyard | Interior courtyard | Exterior courtyard

Kitchen
The open kitchen facing the dining room constructs the warm atmosphere in a small spcae.

Private courtyard
The house part has the private courtyard with dirct skylight. Every room has the bridge to connect with each other.

Staircases
The staircases go upper around the different middle courtyards.

Gallery on the 2/F
It is the extention of gallery on the 1/F and the transition to the more private working space.

可持续漂浮住宅设计
（新南威尔士大学）

新南威尔士建筑学 offer 尹媛

According to the global warming, which leads to more precipitation, the frequency of the snow and rain are becoming higher and higher, more intense is the Precipitation, there are higher risk of floods. That means the Climate change will make this disaster more frequent and more extreme. During the last two years, from Pakistan to the United States, there are many floods, which caused serious economic losses and casualties. In order to reduce the losses in the floods, amphibious houses might be a good choice.

"amphibious houses" – a concept that can make sense when the solid ground is yearly sinking globally. People are looking into the future with concern over floods caused by heavy rain and other disasters, an issue that will be compounded by rising sea levels from global warming. The 'Amphibious Houses' use a prefabricated steel floatable system built over a tank. When the area begins to be flooded, water first pours into the tank and the house begins to float and rise up with the water. The planning of the neighborhood includes vegetable gardens and residential houses, which work together as one floating community.

Backup systems such as rainwater collector, solar power system, and protected food storage areas allow each house to function as normal even if the civil utilities fail. The new houses will be arranged in mini-communities typically made out of 5-10 families, so families can assist each other during the flood until more help arrives. Ideally, as the buildings can float, no property will be lost or damaged and people can continue to living as normal, except for traveling by boat rather than by foot.

方案理念介绍

Nine residential forms provide residents different living possibilities, the wooden plank on the outside wall of the house is able to become a bridge connecting the adjacent blocks when flood comes. This let the residents escape from the form of island.

The vegetable garden can be shared between different families, the vegetables planted in the garden can satisfy part of the daily food consumption when the flood comesand provides the opportunity to communicate during the normal days.

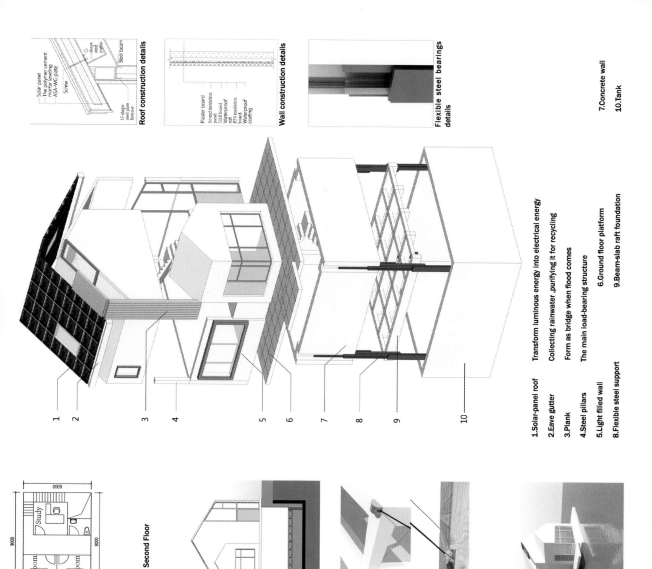

Roof construction details
Wall construction details
Flexible steel bearings details

1. Solar-panel roof — Transform luminous energy into electrical energy
2. Eave gutter — Collecting rainwater, purifying it for recycling
3. Plank — Form as bridge when flood comes
4. Steel pillars — The main load-bearing structure
5. Light filled wall
6. Ground floor platform
7. Concrete wall
8. Flexible steel support
9. Beam-slab raft foundation
10. Tank

Second Floor / Ground Floor / Basement

Floating Process

Bridge construction details

Transforming process of Bridge

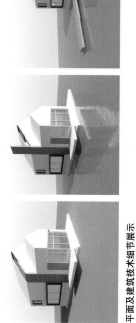

平面反建筑技术细节展示

模块化住宅设计（墨尔本大学）
墨尔本大学建筑学硕士 刘宇恒

The Unfinished Project

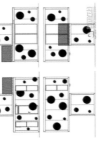  

Advantage of Container House

1. Convenient transportation, especially to the unit that need to change the construction site frequently;
2. It is strong and durable because it is made of steel with strong earthquake resistance and anti-deformation capacity;
3. The mobile house has good water tightness for good sealing and strict manufacturing process

Change container house into living house (a house can be close up)

Press the button and the surrounded walls stretch like flower, then the container is changed into a living house. The house with the container concept is designed by architect, Adam Kalkin. There is full with bedroom, toilet, kitchen and living room. The house will recover to container with one button. It is very convenient for transportation.

1. Environmental protection. Since it adopts to factory production, the construction waste in the construction site is decreased largely. It will be more environmental protection.

2. Panel saving. Since the laminated plates is made as floor slab floor and the outside panel is one side panel of shear wall, large quantity of panel is saved greatly.

| Design Idea

The architecture seeks to realize the evolution of the architecture by the students the building is transited from fixed space around to intermediate flexible space. The surrounding fixed space is designed to arrange the long term fixed function. The students can change the middle flexible part to meet their multiply activities. The flexible frame architecture will make contributing to students' imagination and operation ability.

The flexible space uses flexible and portable maintenance structure. Students are able to move the floor and wall optionally and create the building space which will meet their need.

The Unfinished Project

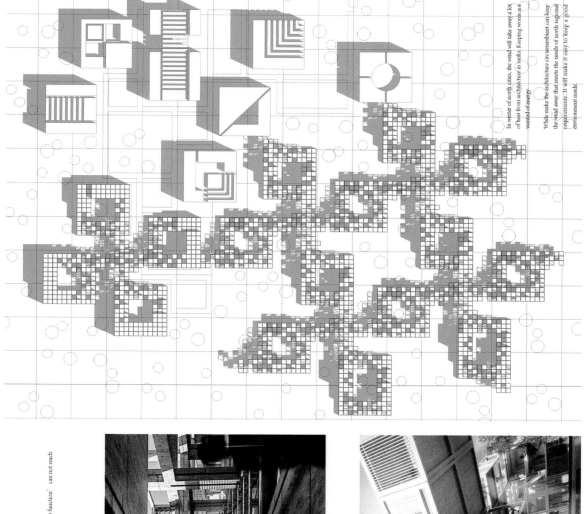

In winter of north cities, the wind will take away a lot of heat from architecture in racks. Keeping warm is a wanted of energy.

While make the architecture circumambient can keep the wind away that meets the needs of north regional requirements. It will make it easy to keep a good environment inside.

Background

With the development of society, the improvement of technology and the demand of people the Background:Fixed model ' form to follows function ' can not reach the changing demand of users. Old ones will be broken down in a short time in order to built new ones to meet the demand.

Rendering: Architecture designing is the entity space providing places for daily education activities.

Rendering: Architecture designing is the frame space which provides a possibility for students to design by themselves.

The Unfinished Project

Design Ideas
If the architect is unfinished, it will not be eliminated. It will have infinite possibilities and high usage rate.

Rendering
From always follows the function, the function remains unchanged, the form is also unchanged.

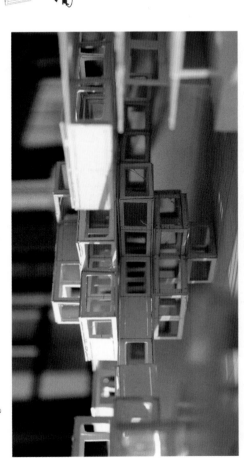

Rendering
It is strong and durable because it is made of steel with strong earthquake resistance and anti-deformation capacity.

Space And Function

- Grey Space
- Private
- Public Space
- Outdoor Space

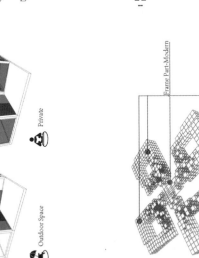

Frame Part-Modern
Entity Part-Tradition

Construction Structure

Roof and Floor
Fixed on crosswise slide pole they can move athwartships.

Flexible Maintenance Structure
Fixed on cross vertical sude pole which can move in the vertical ground.

Column Grid Of 6M
To undertake the weight and provide the vertical pole.

H-Beams
To transferthe load and provide the crosswise slide pole.

Regionalism

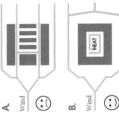

A. Wind
B. Wind

Flexible Evolution

The architecture is more flexiblein the middle part.

The Unfinished Project

Design Ideas

The forum of architecture from the needs of users. As the main users, we students have a great influence on the formation of buildings. As architecture students we emphasize the importance of hands-on practice all the time. In the design process, we need to use the manual moodel to study the body, express ideas, and understand the building through the actual operation. Structure which makes us fear the actual projects. There comes the building festival in some colleges using the cardboard to build what you want in order to use the design work in person.

Rendering: Architecture designing is the entity space providing places for daily education activities.

Rendering People can do different activities in the space they create and change the space form according to function demand.

1. Roof Panel
2. Side Wall Panel
3. Recycled Shipping
4. Connections
5. Plywood Siding
6. Wire Rack
7. Corner Post
8. Thermal insulation
9. Joist Steel
10. Front Doors
11. Sheet Metal Panels
12. Bottom Panel

Energy saving.

Because the outside panel is that 50 plastic extruded board is caught in the middle of concrete with two sides to make the thermal insulation property is better than the property of the outer or inner the outside wall of traditional architecture. Meanwhile, it also solve the outside wall decoration off phenomenon caused by the outer thermal insulation property in traditional architecture.

茶室花房类设计作品
Tea house/Green house category

花房设计（米兰理工大学）
米兰理工大学建筑设计专业　耿直

本设计选取中国北方住宅小区中闲置的绿化空地作为设计场地，设计灵活可变的花房满足社区居民不同季节的社交活动需求。

SMALL COMMUNITY CENTER AND GREEN HOUSE

(second semester in the fourth year of college, individual work)

Located in Mudanjiang City, Heilongjiang Province, the northernmost part of China. Five months of a year below 0°C. It is a parterre spared in autumn and winter

KEYWORD:
fixed and changed
planting circle
ice gable

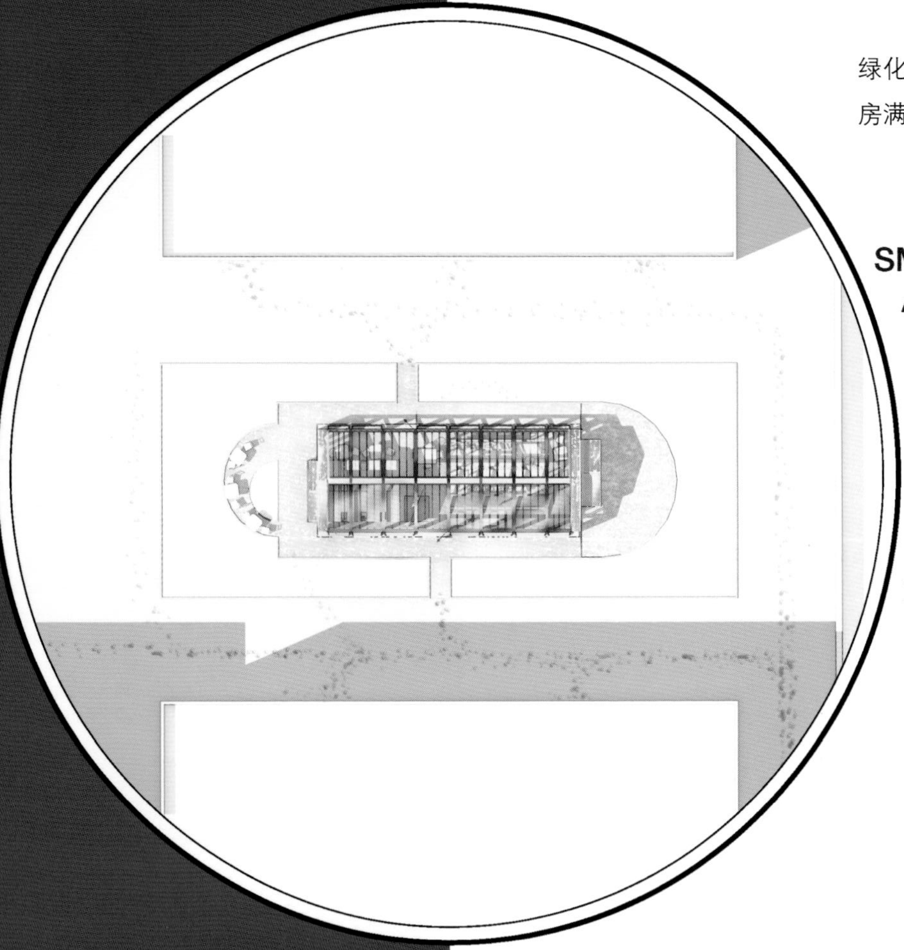

方案背景介绍

WINTER

SUMMER

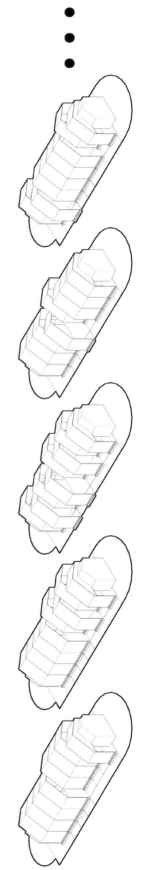

冬夏设计转化展示图

茶室花房类设计作品
Tea house/Green house category

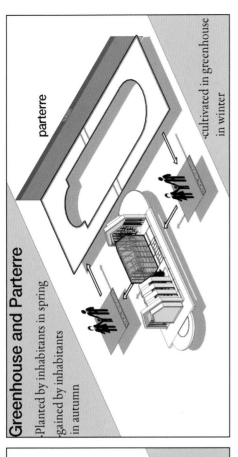

Greenhouse and Parterre
- Planted by inhabitants in spring
- gained by inhabitants in autumn
- cultivated in greenhouse in winter

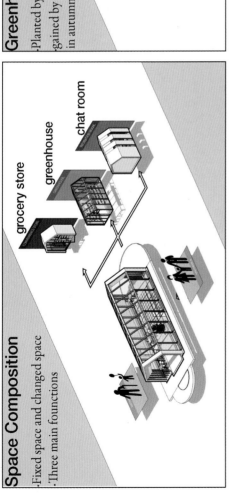

Space Composition
- Fixed space and changed space
- Three main founctions

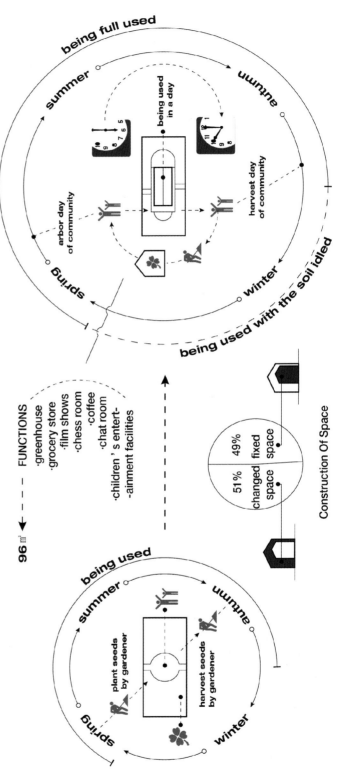

设计理念详解图

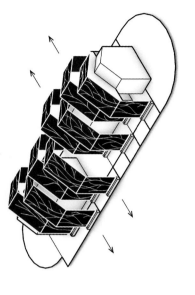

SPRING

Plant the creepers in the planting box

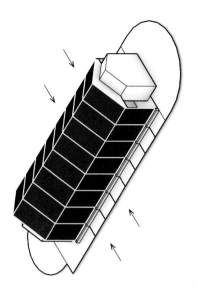

SUMMER

Open the skin to get better ventilation, and the grown-up creepers can replace the sun visor

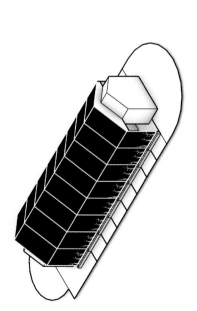

WINTER

Build the gable with bricks made of ice to against −30°C The insolation of ice is much better than concrete especially wih air

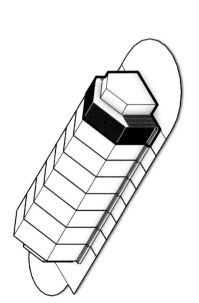

AUTUMN

Close the skin to keep warm from the cold wind. More sunlight is able to get into the house because of the wilting of the creepers

四季轮回的变化图

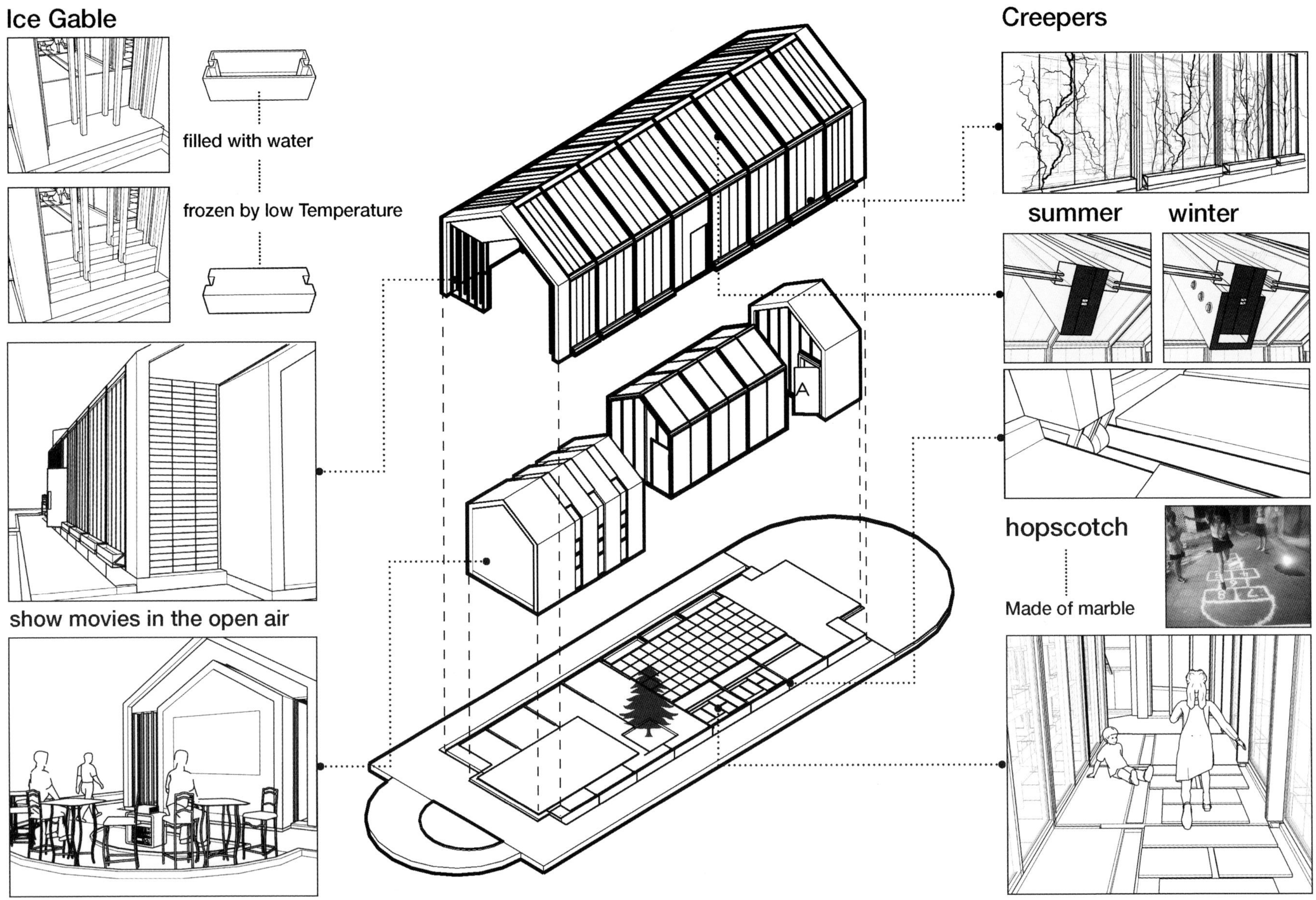

Creepers

winter

summer

hopscotch

Made of marble

Ice Gable

filled with water

frozen by low Temperature

show movies in the open air

构造细节与场景设计展示图

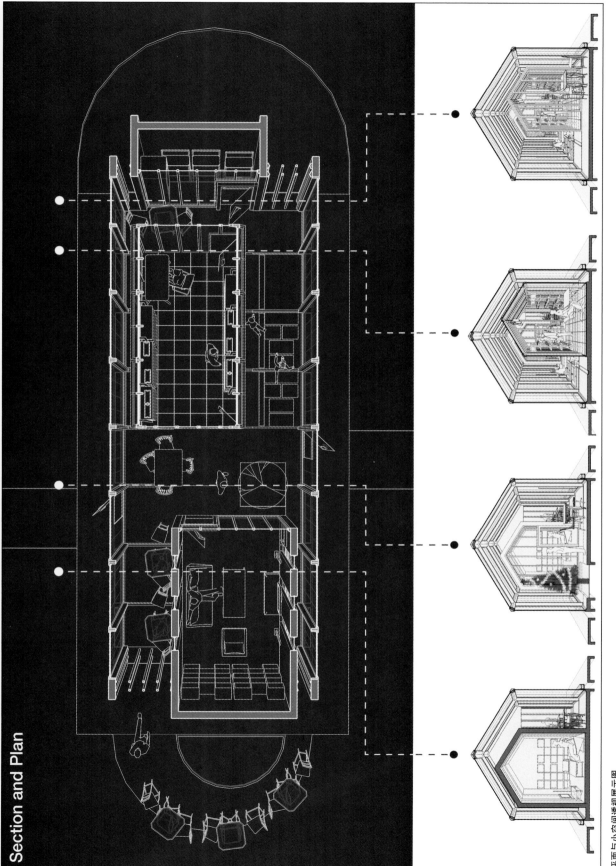

平面与小空间透视图示图

茶室设计（谢菲尔德大学）

谢菲尔德大学建筑学硕士　李怡萱

THE TEA ROOM

Time:Autumn Semester~2012(sophomore year)
Duration:Two Weeks
Location:The West Lake,Hangzhou,China
Critics:Bing Li,Sister Mai

People are as busy as a bee in weekdays in cities,especially office workers,they are under too much pressure in daily life.So they are eager to go away from the noise city to beathe fresh air and relax in holidays.But there are many people in scenic spots in holidays,and there are few place for tourists to take a rst.People are surrounded by a crowd so that they can not to really relax their heart and body.

The aim of the tea room here is to give tourists a shelter where they can have a rest,drink tea and enjoy the scenery.And people can medicate here to think about their lives.

Location Instruction:

This project is located in Hangzhou province.It is in the north-west of the West Lake,which is the one of the most beautiful lakes in China. The scenery of the West Lake is different in four seasons.In spring,all the plants are sprouting,everything wake up.In summer,the water lilies are opened,andthere are so much rain that we can enjoy the rainy day of this lake. In autumn,we can see the beauty of falling leaves.In winter,we can also playaroud the lake,because of the warm climate in Hangzhou. This project is a natural place for people to enjoy the fragrance of tea.

Sloping Roof

Footpath

Pavilion

Enframed Scenery

Lattice Window

理念生成阐述及场地分析图

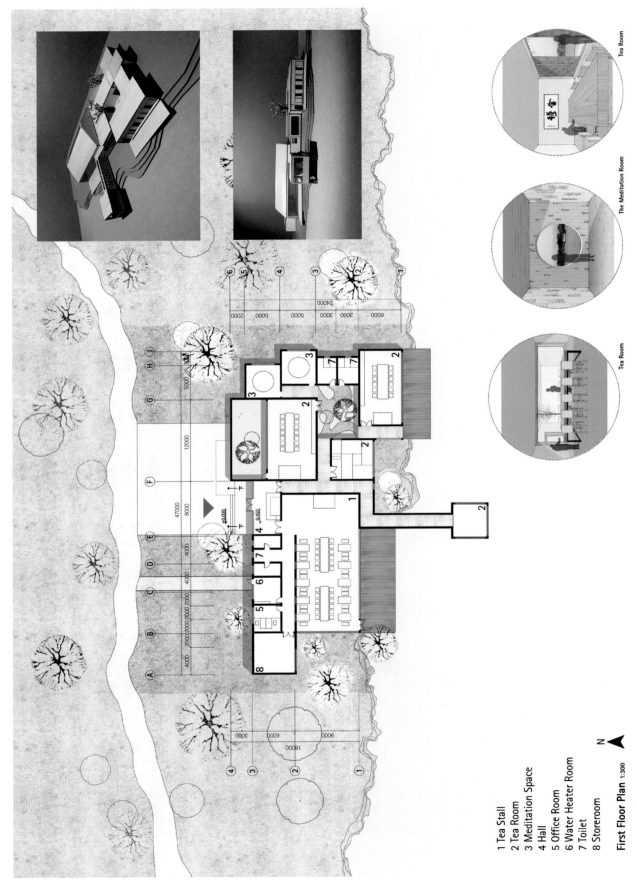

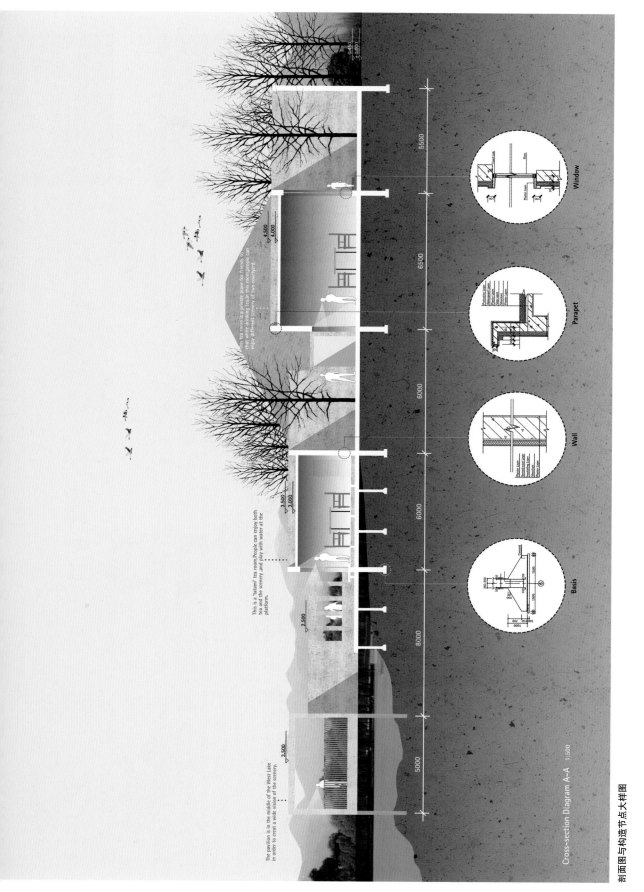

Design Process:

This project starts as a cuboid. According to its function, we divide this cuboid into two parts and dug two holes on it to make courtyards. Then, we adjust the hight of one part to provide premise conditions to create a private room. Because of adjoining to the West Lake, the south side retreat a few to make sure the sight of the West Lake. Then, we subarea the low hight part into three tea rooms, two meditation rooms and a toilet. At last, we creat two water platforms for tourists to be close to nature.

Tea Stall:

It is a place that tourists have a rest and have some tea. There are a row of French windows in the south of tea stall. People there can both drink tea and enjoy the scenery of the West Lake.

Tea Room:

There are three different kinds of tea rooms in the east part of this building. The first tea room is a square 'tatami' room where people sitting there can enjoy the beautiful scenery of the West Lake. The second one is a private space in the face of a introverted courtyard. The third one is a half-open space with a waterborne platform. These tea rooms are private part of this building.

Tea Meditation Space:

There are two meditation spaces. They provides places for people who are busy in cities to deep thought. These spaces with wide field of vision are at the southeast corner of this building in order to creat a quiet atmosphere.

Pavilion in the lake:

The pavilion is in the West Lake, and it is a place with wide vision for people to enjoy drinking tea and seeing the scenery. This pavision is surrounded by water. There is a narrow path for people to get there to ensure the quiet.

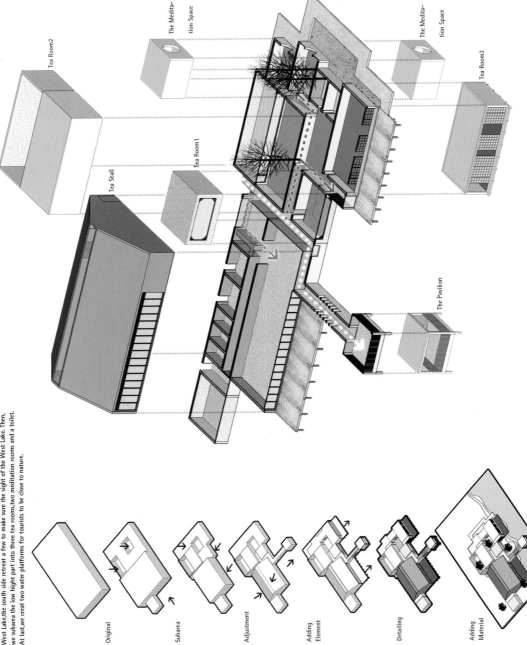

Tourists Line
VIP Line
Viewing Line

Exploded Drawing

Original
Subarea
Adjustment
Adding Element
Detailing
Adding Material
Result

Tea Room2
Tea Stall
Tea Room1
The Meditation Space
The Meditation Space
Tea Room3
The Pavilion

Space Sentiment Analysis

This project is eager to change people's sentiments through different kinds of spaces,In tea stall,people can relax and enjoy the whole scale scenery of the West Lake.In tea rooms, people can drink tea and see apart of this lake.In the meditation rooms,people will be quiet to think about their lives.

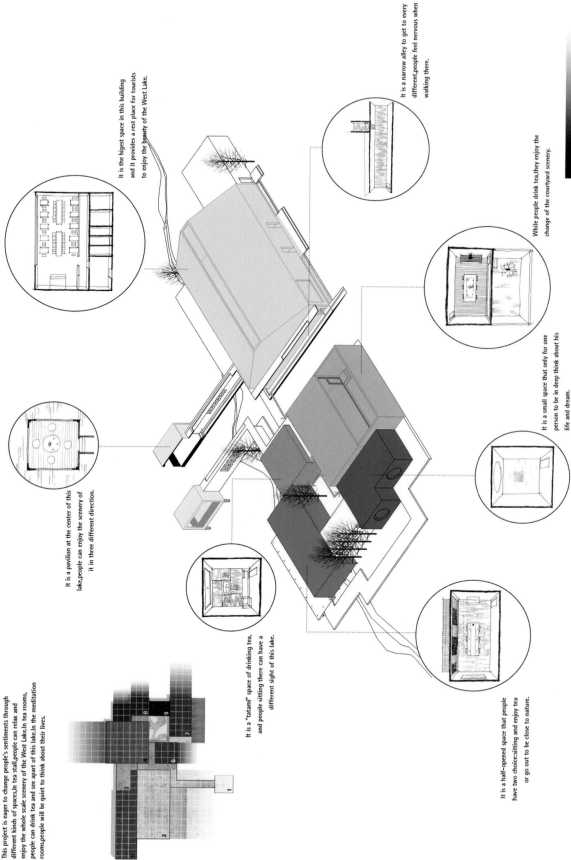

It is a pavilion at the center of this lake,people can enjoy the scenery of it in three different direction.

It is the bigest space in this building and it provides a rest place for tourists to enjoy the beauty of the West Lake.

It is a narrow alley to get to every different,people feel nervous when walking there.

While people drink tea,they enjoy the change of the courtyard scenery.

It is a small space that only for one person to be in deep think about his life and dream.

It is a "tatami" space of drinking tea, and people sitting there can have a different sight of this lake.

It is a half-opened space that people have two choice:sitting and enjoy tea or go out to be close to nature.

Open

Private

功能细分介绍图

KINDERGARTEN -'L' shape

幼儿园及小学类建筑设计作品

香港 L-shape 幼儿园设计（香港中文大学）

香港中文大学建筑学硕士

朱启迪

场地坐落在香港 Aldrich Bay 公园，有一定的地形高差变化。该设计旨在创造一个充满安全感的幼儿活动学习空间。同时通过丰富多变的室内外空间和充分的自然采光，露合和天桥的变化，将狭窄场地的幼儿园变成一个充满童趣、明亮而丰富多彩的小院落。

建筑形体的错落变化，自然采光的巧妙设计以及活动空间的多样化是本设计的亮点。建筑与地形高差的自然结合也体现了设计者对环境的充分考虑。

Finished time: June 2015
During the second year of undergraduate program of Architecture design and theory
Individual work
Location: Aldrich Bay Park, Hong Kong

建筑入口透视图

幼儿园及小学类建筑设计作品
Kindergarten/Primary School category

Design theory

The kindergarten is designed in three floors on a gentle slope. Because the site is in the park, it has great views of landscape. I first use the 'L' shape block to define the open space by putting the 'L' shape in different directions, rotating it and cutting out the spare part from the rectangular blocks. For the function, the lower two floors are classrooms and play areas, the top floor is the gallery and waiting area for parents.

Master layout plan

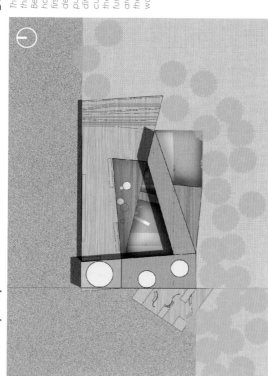

Concept diagram

1. What happens to the vacant area in the park?

2. Taking out all the soil of the slope.

3. Building a box.

4. Cutting out the spare block and forming the 'L' shape.

5. Cutting out the spare blocks at every layers from different directions.

6. Rotating the shape to have more opportunities.

Outdoor play area

The main play area is on the ground floor in the middle courtyard. The run track passing through the indoor and outdoor play area provides enough active area for children. The children on the first floor can use the slide directly to the ground courtyard.

方案理念及场景透视图

Indoor play area

The indoor play area is next to the outdoor play area by using the thick wall to separate. There are holes in different size on the wall providing for children to climb.

Gallery

The gallery on the second floor shows children's art works and drawings. Circle windows on the roof allow the direct skylight into gallery.

Plans

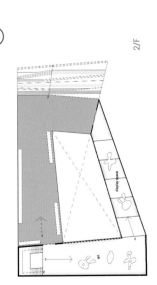

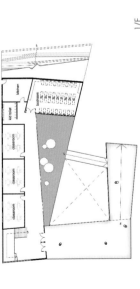
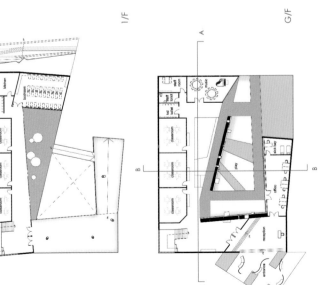

幼儿园平面图及游戏空间透视图

Second entrance
Parents can wait for their children at this second entrance in the park. Public can also sit on this staircase and enjoy the great view of park.

Park beside the kindergarten
This kindergarten locates in the park. After school, a lot of activities happen here, such as playing football, flying the kite and walking the dog and etc...

Section B

Classroom | Indoor play area | Outdoor play area

Section A

Bedroom | Indoor play area | Gallery | Main entrance

室外场地及剖面图

小学设计（谢菲尔德大学）
谢菲尔德大学建筑学硕士 李怡萱

地点：中国山东青岛市。

本设计旨在为小学生创造出学习和玩耍共融的新学校建筑空间。

入口透视效果图

Primary School

Time:Autumn Semester-2013(junior year)
Duration:Three Weeks
Location:Huangdao,Qingdao,China
Critics:Tong Zhou,Sister Mai

When children go to the primary school,they stay in classroom almost the whole day,the out-of-class time for children is short,but the distance between classroom and activity place is far.So that they spend more time running between these two places,they just have a little time playing,and children are unsafe to running.

108

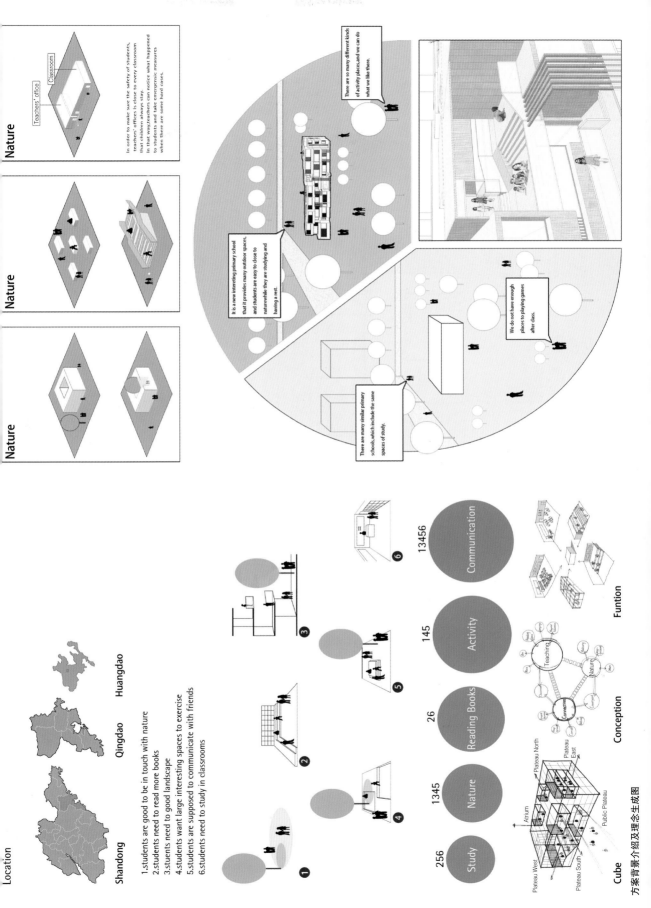

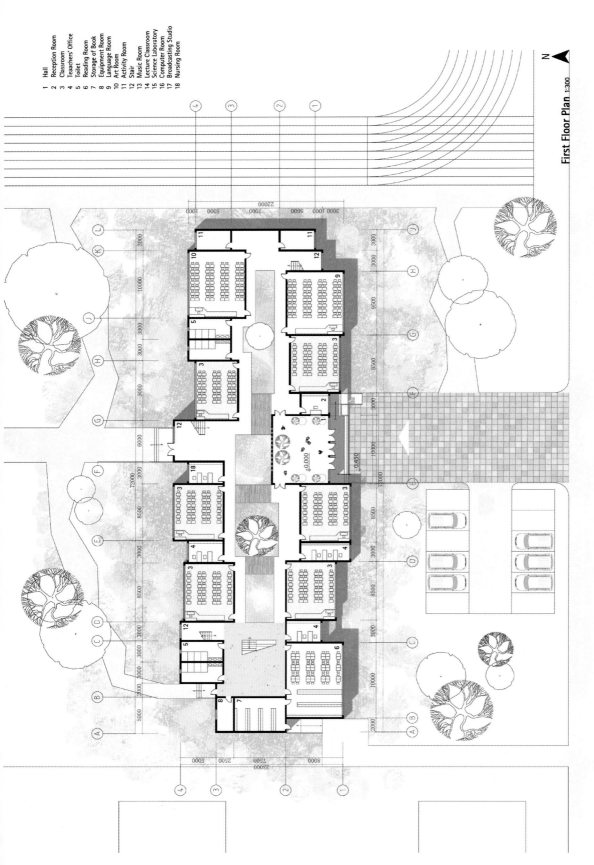

First Floor Plan 1:300

1 Hall
2 Reception Room
3 Classroom
4 Teachers' Office
5 Toilet
6 Reading Room
7 Storage of Book
8 Equipment Room
9 Language Room
10 Art Room
11 Activity Room
12 Stair
13 Music Room
14 Lecture Classroom
15 Science Laboratory
16 Computer Room
17 Broadcasting Studio
18 Nursing Room

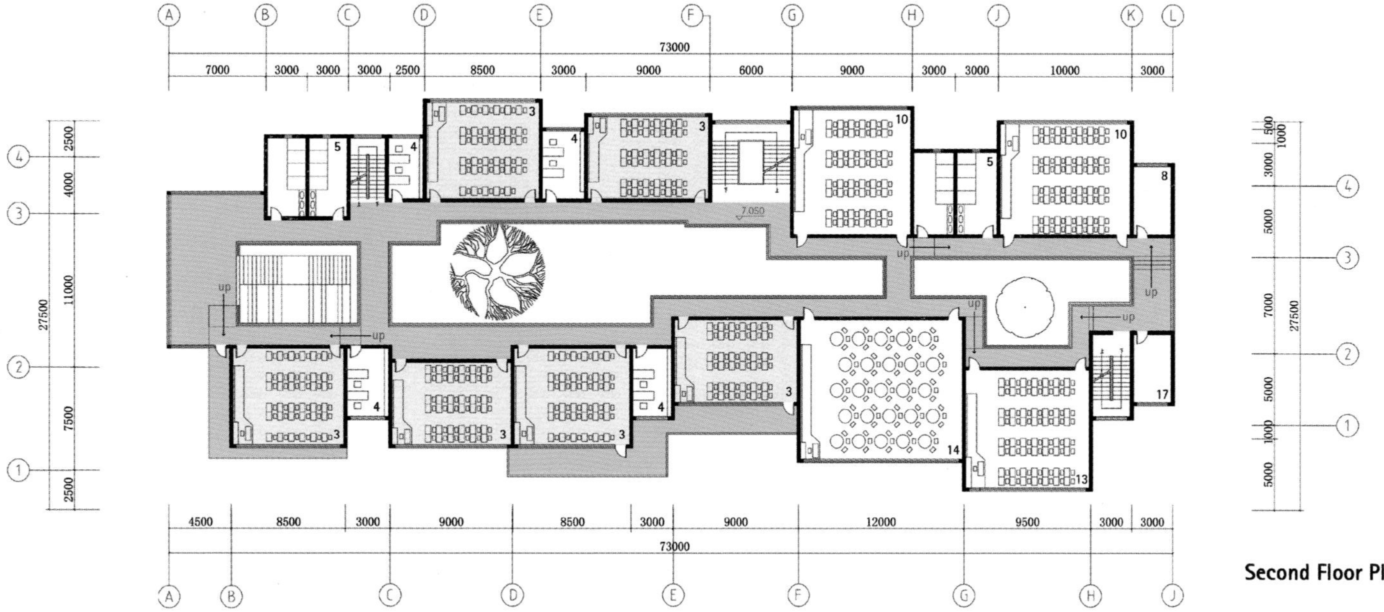

Second Floor Plan 1: 300

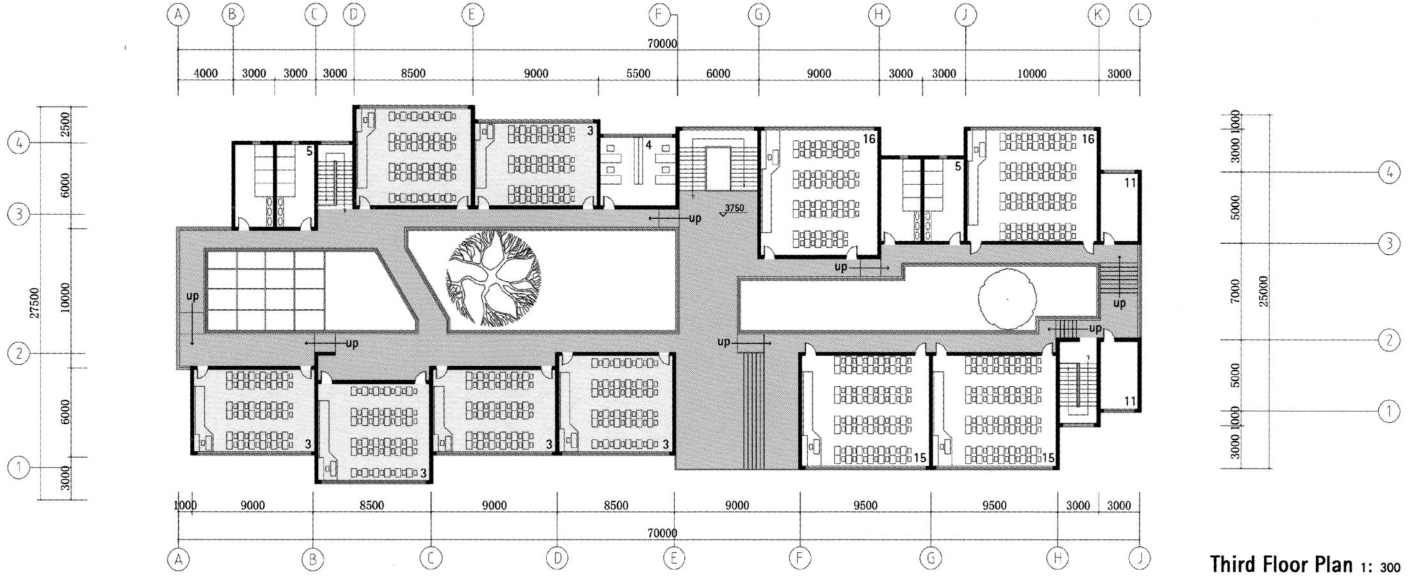

Third Floor Plan 1: 300

二、三层建筑平面图

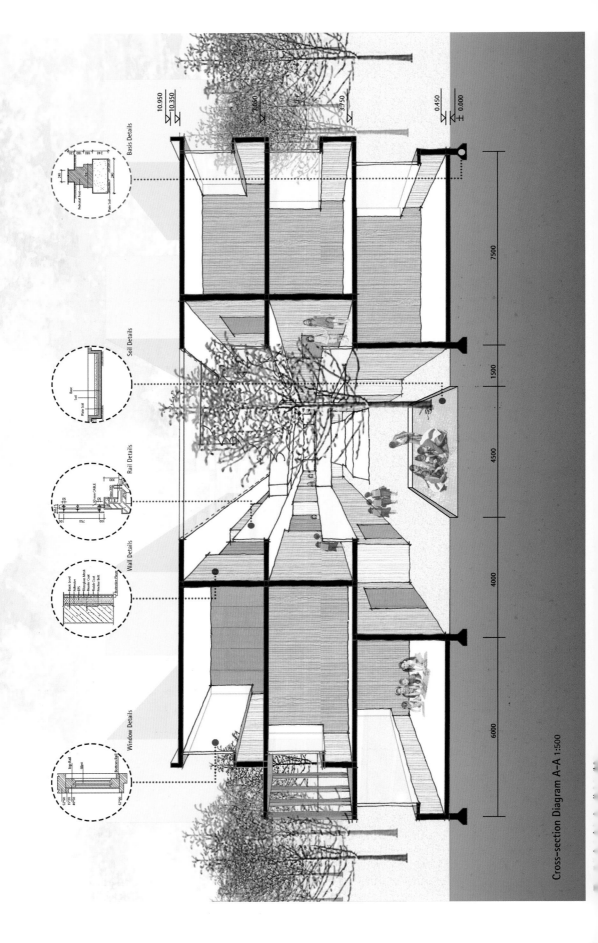

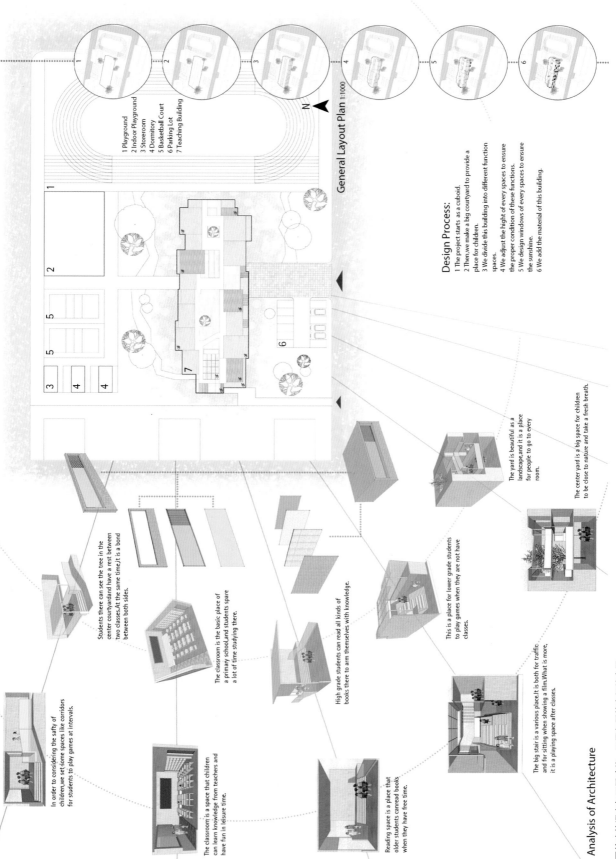

Breakdown Drawing 1:500

First Picture:
- It is a front view of this building,people enter into it from the passageway of the frontage.
- It is the main side of the building that it shows the differ -ences of every rooms.
- There is a large square in front of the building for cars parking and people walking.

Second Picture:
- It is a corner of this building.
- There is a playform in the second floor of this building that students can playing at intervals.
- There are some residental building besides this primary school.
- There is a footpath in the west of this building that peo -ple can go to the dormitories throuth it .

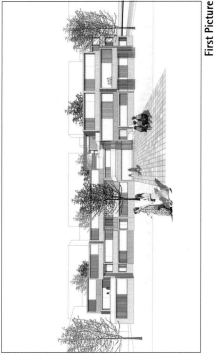

First Picture

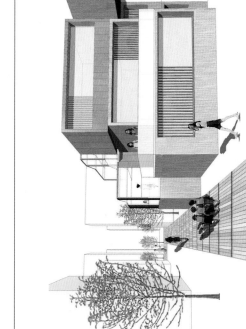

Second Picture

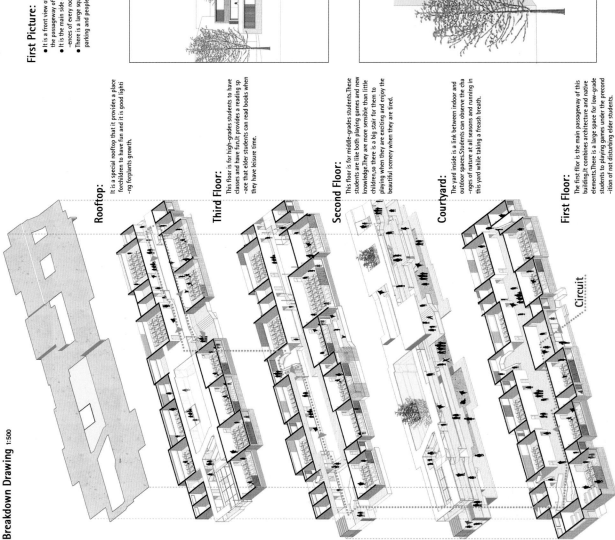

Rooftop:
It is a special rooftop that it provides a place forchildren to have fun and it is good lighti -ng forplants growth.

Third Floor:
This floor is for high-grades students to have classes and have fun.It provides a reading sp -ace that elder students can read books when they have leisure time.

Second Floor:
This floor is for middle-grades students.These students are like both playing games and new knowledge.They are more sensible than little children,so there is a big stair for them to playing when they are exciting and enjoy the beautiful scenery when they are tired.

Courtyard:
The yard inside is a link between indoor and outdoor spaces.Students can observe the cha -nges of nature at all seasons and running in this yard while taking a freash breath.

First Floor:
The first flor is the main passageway of this building.It combines architecture and native elements.There is a large space for low-grade students to playing games under the precond -ition of not disturbing elder students.

Circuit

Some photos of material

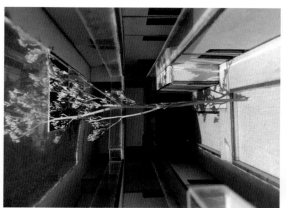

整体与细节模型展示丰富的室内外空间变化图

博物馆展览馆类建筑设计
Museum/Gallery category

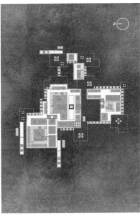

MASTER PLAN

Concepts

Background

实体模型及方案理念阐述图

窑洞博物馆设计（瑞典皇家理工学院）

瑞典皇家理工学院建筑学硕士 吴月

方案以中国西北传统民居窑洞为出发点，通过提炼和再表达传统的下沉式窑洞建筑语言，将博物馆展览空间和中国传统围合院落以及园林空间过渡衔接的手法结合在一起，在宽阔的黄土高坡上设计了一处只闻人语声的地下博物馆。虽然采用整体下沉设计，但是整个博物馆的采光设计良好，不同层高的院落可以给参观者带来步移易景的视觉效果。

PLANS

1st Basement floor

1. Entrance hall
2. Service office
3. Elevator room
4. W.C.
5. Exibition room
6. Storage room
7. Multi-function room
8. Experience room
9. Meeting room
10. Cantine
11. Scenery viewing tower
12. Service courtyard
13. Courtyard

2nd Basement Floor

1. Exibition hall
2. Storage room
3. Souvenir shop
4. Storage room
5. Elevator room
6. W.C.
7. Experience room
8. Scenery viewing tower
9. Courtyard

SECTION PLANS

1-1 SECTION PLAN

2-2 SECTION PLAN

3-3 SECTION PLAN

4-4 SECTION PLAN

5-5 SECTION PLAN

6-6 SECTION PLAN

7-7 SECTION PLAN

剖面及建筑主要平面图

博物馆展览馆类建筑设计
Museum/Gallery category

ART OF SCENERY CREATING OF CHINESE GARDEN

This underground museum design integrated the cave dwelling culture and the traditional Chinese garden art together. Scenery-creating arts are used in the courtyards, generating interesting relationship between interior and exterior environment. When tourists walking through the route, the scenery will change along the route step by step.

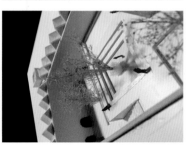

The underground museum integrated the dark and tranquil cave atmosphere of cave and various sceneries of Chinese garden together, creating a series of attracting courtyards and amazing interaction space between interior and exterior.

The sunken courtyards are located at different altitude where tourists can experience the amazing spaces at different height, the rising tower offers great view seeing opportunity of the whole museum while water landscape surrounding the tower promotes the vitality of the museum.

室内场景渲染和局部空间模型

厦门大学科学博物馆设计
（米兰理工大学）
米兰理工大学建筑学硕士
李炫静

设计紧邻厦门大学，有便捷的交通和优美的水景观。本方案将建筑与场地交织设计，将建筑开敞的抬高落与场地串联在一起。通过游览式的院落设计将建筑和周围的校园环境巧妙地融合在一起。

Period : July, 2016
Academic Year : Fourth
Location : Xiamen City, Fujian Province, China
Site Area : 11,835 m²
Type : Project Assignment | Individual Work

04 SCIENCE MUSEUM DESIGN

The site locates in Xiamen university, Fujian province, China. It adjacent to the main roads of the campus with large flow of people. There is a beautiful lake on the north side and a number of teaching buildings separately on the west side and south side.

In recent years, with the growing number of visitors to the campus and the development of university research, the university needs a science museum to show its scientific achievements and the strong academic atmosphere to the outside world, also an essential place for students to communicate, learn and relax. Therefore, the science museum should contain the spaces for display, comprehensive research and communication (library, academic lecture hall, etc), administrative office, cultural relics storage, public services and other functions.

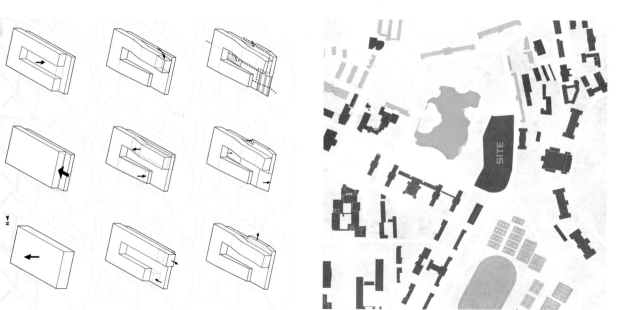

- - - Road System
- Teaching Building
- Life Service Zone
- Dormitory
- Sports Ground
- Lake

基地分析与形体生成展示图

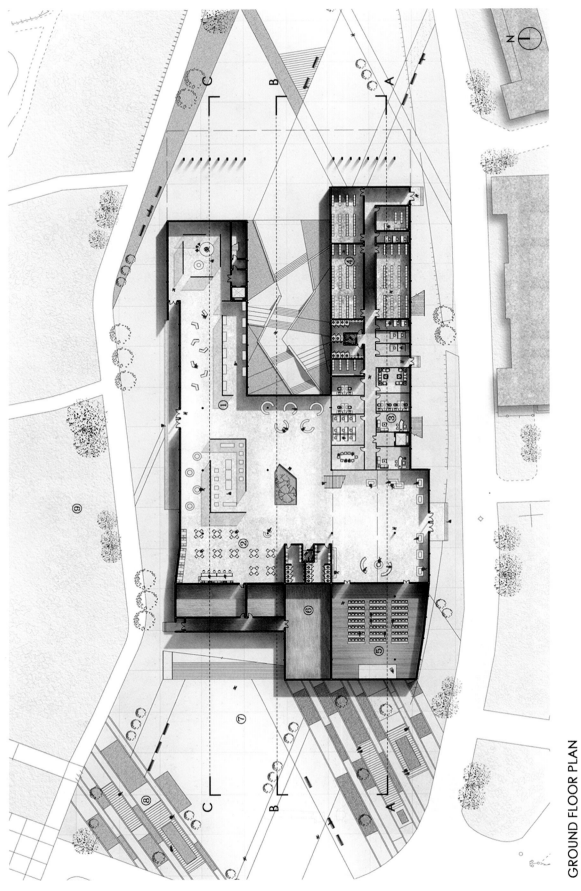

GROUND FLOOR PLAN

① Exhibition Area ② Restaurant ③ Office
④ Storage room ⑤ Academic Hall ⑥ Auxiliary Function Room
⑦ Entrance ⑧ Landscape ⑨ Green

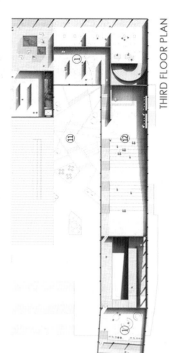

PERSPECTIVE SECTION
THIRD FLOOR PLAN
SECOND FLOOR PLAN

① Exhibition Area
③ Office
⑩ Cafe
⑪ Courtyard
⑫ Reading Room

剖透图与二三层建筑平面图

博物馆展览馆类建筑设计
Museum/Gallery category

There is a courtyard between the "U"-shaped building with a changing-width path going down crossing the landscape, connecting the cafe and show spaces on both sides of the museum. Also, the courtyard provides visitors and students with an excellent outdoor recreation space, as well as a landscape resource within the building.

COURTYARD

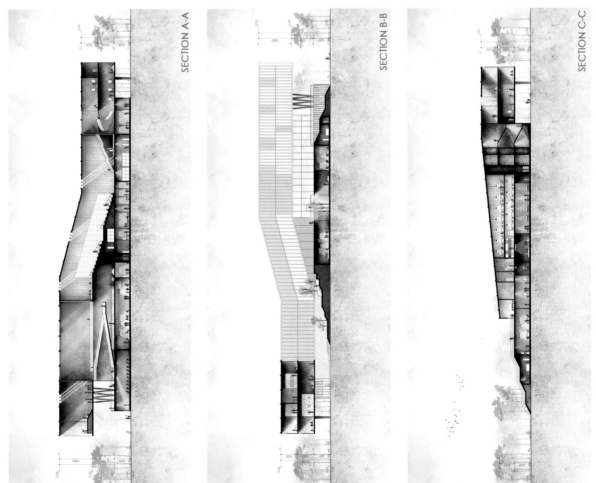

SECTION A-A

SECTION B-B

SECTION C-C

剖面图与建筑庭院透视图

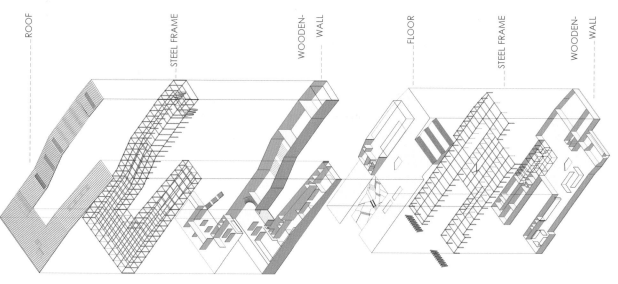

PERSPECTIVE

With the growing number of visitors to the campus and the development of university research, the university needs a science museum to show its scientific achievements and the strong academic atmosphere to the outside world, also an essential place for students to communicate, learn and relax. Therefore, the science museum should contain the spaces for display, comprehensive research and communication (library, academic lecture hall, etc.), administrative office, cultural relics storage, public services and other functions.

广场透视图与结构爆炸图

博物馆展览馆类建筑设计
Museum/Gallery category

READING ROOM

Reading room gives students a space to communicate and read, with a dedicated entrance. Visitors would also be here and experience the university's academic atmosphere.

阶梯教室渲染图

展厅室内渲染图

The Museum of Memory
Time:Spring Semester-2014(junior year)
Duration:Four Weeks
Location:Kashgar,Xinjiang,China
Critics:Jianfeng Xu.

There are many beautiful memories in people's minds,including trivial affairs,touching love stories and memories of their childhood.According to share these memories to more people,we creat a memory museum toprovide a place for them who miss the old days,like reading stories to recall old things in their minds and enjoy the good memories,whatever theirs or others',then,they will be happier to live better.

This memory museum consists of three places for people to recall these old memories,including story rooms,memory corridor and the exhibition of time.These places are a long way to bring people back to the past of their own memory and experience their stories in their minds.What is more,we put some medication boxes in the sqaure for people who have visited this museum to thinking their lives.

记忆博物馆设计（谢菲尔德大学）
谢菲尔德大学建筑学硕士　李怡萱

本设计根据记忆的特点设计了独特的记忆博物馆，包括故事房间、记忆走廊和时间展厅。

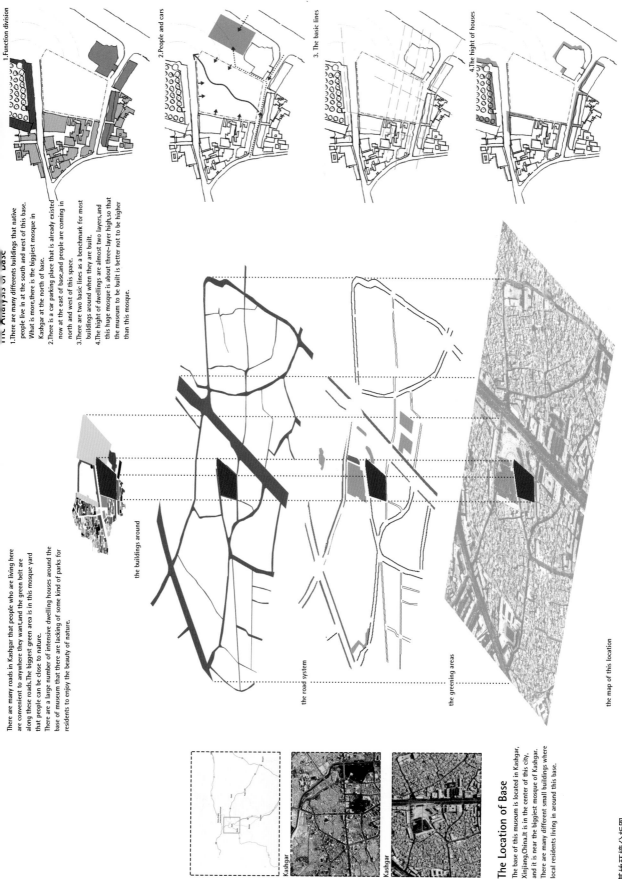

The Conception of Interior and Outerior

1. There will be some huge stairs for people to go to the rooms underground. These stairs outsides are a kind of place for them to talk to others and have a rest.

2. There will be some courtyards planting trees for people to be close to nature and to enjoy the beauty of nature. What is more, the sunlight of rooms under the ground are from the windows on the ceilings.

3. There will be some stairs indoors to line these rooms for exhibition, and there are some small yard for sunlight of rooms underground.

4. There will be a multifunctional hall for people to get together to do activities. On the ceiling of this hall there is a large stairs for people to sit down to relax and chat with friends.

The Ethic Elements

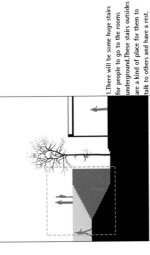
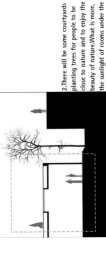
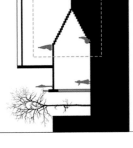
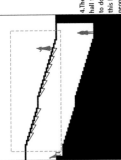

1. Lattice Window:
There are a series of windows with arch at the wall of corridors.

2. Islam Style:
The islam style are common in Sinkiang and there always be some fire door opening coupons arch-shaped.

3. Lattice Door:
There are some lattice doors with beautiful patterns of Uyghur nationality.

4. Fire Door:
The doors there are arch-shaped at the top of them and patterns on doors are different with different means.

5. Wooden Constructions:
As for the special climate, there are many wooden constructions for people to enjoy the cool and hang food out.

Expression Meathods of Houses

There are many courtyards in the houses that local residents are living in, and corridors is always with these courtyards.

Local residents are always using the balconies to dry clothes and food in the sun, and people are playing at the balconies.

There are many small square windows on the walls of houses in order to ensure the ventilation quantity.

There are many houses called 'Ayiwang' around this city that local residents living in. These houses are consist of courtyard for having a relax, two layers rooms for daily life and balcoies for hanging clothes out. The windows of houses are small to adapt the high winds there and the houstop are flat roof because of little rain all year round.

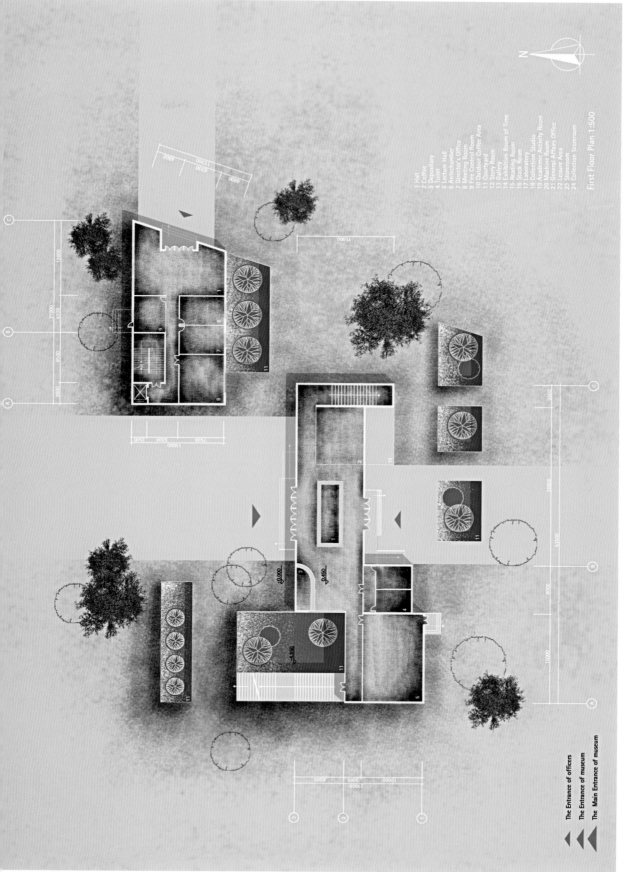

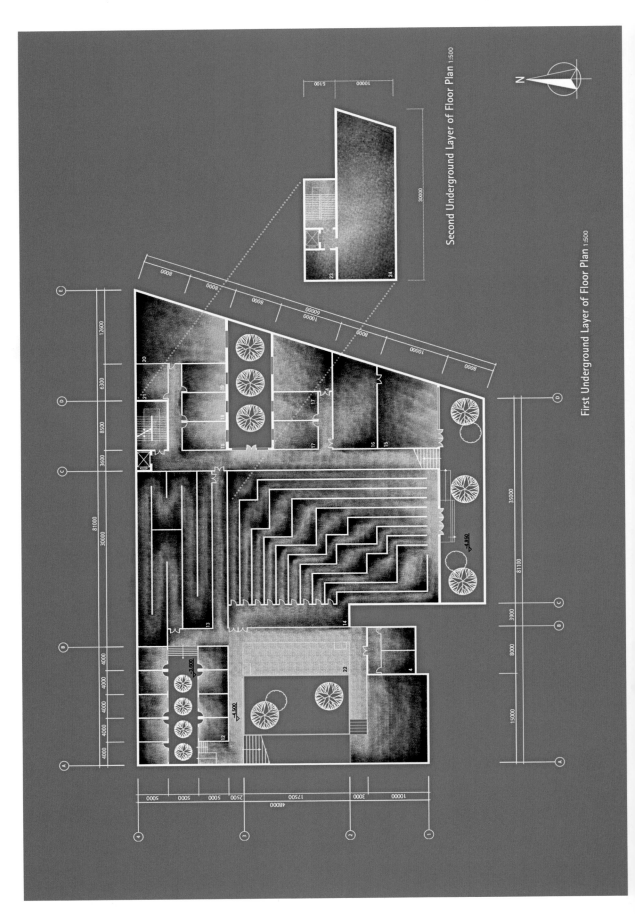

Cross-section Diagram A-A 1:500

Cross-section Diagram B-B 1:500

····· The analysis of Ventilation

建筑剖面图

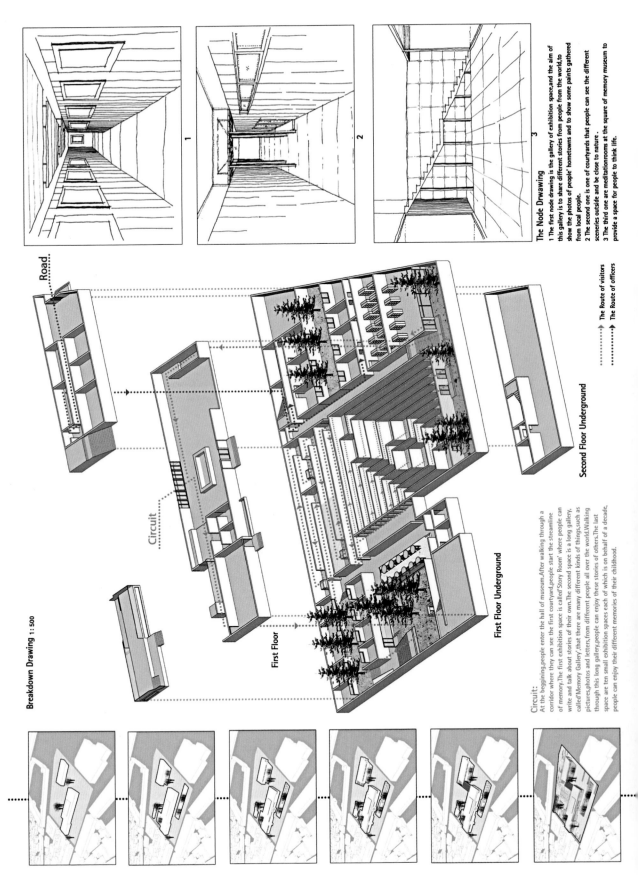

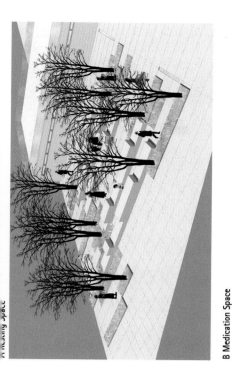

A Resting Space

B Medication Space

C Platform Space

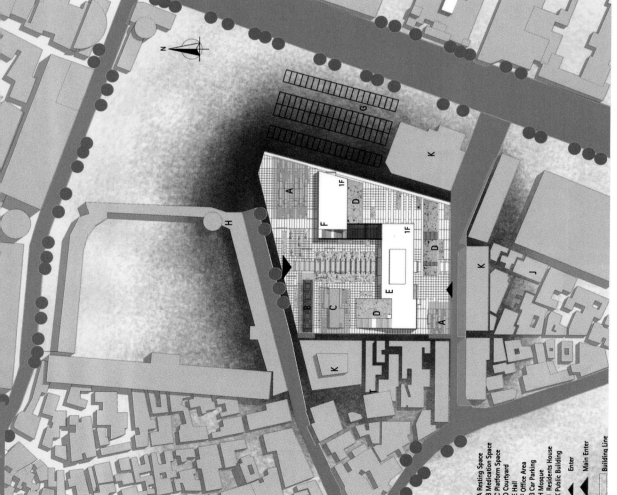

A Resting Space
B Medication Space
C Platform Space
D Courtyard
E Hall
F Office Area
G Car Parking
H Mosque
J Residents House
K Public Building

◄ Enter
◀ Main Enter
▢ Building Line

场地总平面图与小透视图

Introduction

Qianhai D.C. is a newlydeveloped urnban zone in Shenzhen, China. Before the exploit plan of the area had been decided by the urban planner, Qianhai was a small fishing village with its distinctive both natural and cultural features. Thus, with the rapid development, the area will face the following problems,

1. Existing features are dying

According to the research, the fishing culture is dying and the aborigines are forced to move out, this will cause severe cultural vacuum to the area which will be adverse to the future development and regional identifiability.

However, it is inevitable to dismiss the urbanization process cause the ever-increasing materialistic demand of the public. The existing lifestyle and economical structure are of low effectiveness, so the replacement is irrevocable.

Therefore designing a existing culture container will be on the top of our agenda, and designing an architecture combining the usage of virtual reality technology will be a good choice for us.

2. The feasibillity of using VR

Before we started this project, we had a profound research on the technology of VR, because we need to collect the data of the native living styles and the developing process of the city, the VR cameras must have weather resistance and outstanding stability. Finally we found that the device is strong enough.

Time Line

Year 1573
The Birth of Shenzhen.

Year 1978
Chinese Reform caused great changes in the city.

Year 2013
The city lost memories of its past while its economy boomed. Qianhai D.C. was newly planned.

Future
Villages and vernacular buildings of Qianhai will be pulled down. Measures msut have taken to prevent the city from becoming a strange place

Concept

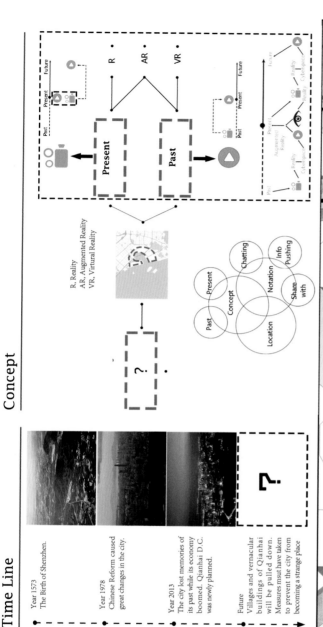

R. Reality
AR, Augmented Reality
VR, Virtual Reality

Site Plan

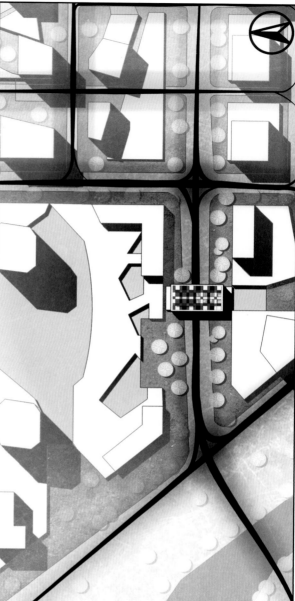

1:3000

This museum is based on the situation that the area is under an enormous change. Thus we intriduce a new concept of using visual technologu such as **VIRTURAL REALITY** and **AUGMENTED REALITY** to compare with present view. Also we introduce a new notaion to fit in this new type of museum, that is the structure of **WECHAT**, defining the features of this software and analogy to the space of the museum. Finally we combine all of them together and a new museum is generated.

博物馆展览馆设计作品

Exploded View

Roof
Roof windows provide vertical eye contact and vertical ventilation.

Exhibition Area
The area provides various routes for visitors to choose, and the exhibition boxes are the place where we compare the difference between past and present by using VR and AR.

General Asile
This element is the "INDEX" of the museum which can guide the visitors to all the different part of the museum by connecting to every regional center

Carrier
The carrier of Qianhai Substitution is a pedestrian bridge of the Qianhai D.C.. This type of museum can be duplicated to different places.

Space Logic

The upwards stairs represents that the visitor surly recognizes this particular show and willing to choose this way will lead them to something relates to the previous exhibition.

The downwards stairs represents that visitors do not really interested in what this particular exhibition and choosing these ways will lead them to change other approach or quit.

A special type of room mainly provide space for visitors to have brief chat no matter what the situation is. This is a space of permeability and a little installation.

This is the mixture of CHATTING and INFO PUSHING, REGIONAL CENTER. The room sets up chiefly serves the connection between different exhibition halls and other types of space. All the cross choice will connect to this area. It is the main participants of the choice movement.

This space is the combination of CHATTING and SHARE WITH. This space particularly provides some spaces for visitors to have a comparatively small scale of sharing circle, similar to private sharing facility.

Whenever visitors see something interesting, ii is easy to record and share it in this particular space by LAN (local area network) of the museum. Everyone's sharing will focus on in this area, as if this area is another exhibition place. Everyone has access to see all the sharing stuffs.

This space combines SHARE WITH and INFO PUSHING. The room primarily represents chatting room of groups of visitors with a bond of sympathy after visit. The space sited near the room of moments for the purpose of building up connections between people with easy access.

Most significant space of the museum that mainly serves the exhibition part. High spontaneity of movement due to the visitors' interest. Fully consider about the response of each individual in the museum. This area uses the technology of VR (virtual reality). People will see the temporary scenery of the city and see what was going on in the past by VR. This provides a straightforward comparison between now and past.

This is the office district for the sake of offering room for the staffs and other administration task. This is a space of good accessibility thus the staffs can easily reach different area of museum.

Functions

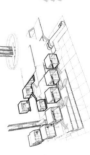

The major exhibition halls divide into four parts based on the four scenery sections during the city development.

Regional Center acts as the HUB of the museum. It connects all the different parts of the museum, and creating buffer space for visitors. All these platforms attached to the "Moments" based on multi-functions.

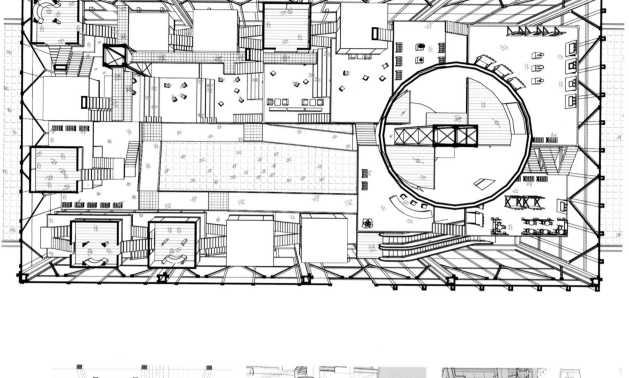

Interior Intention map

I Propose several activities of possibilities for the interior due to the fact of flexible inner space.

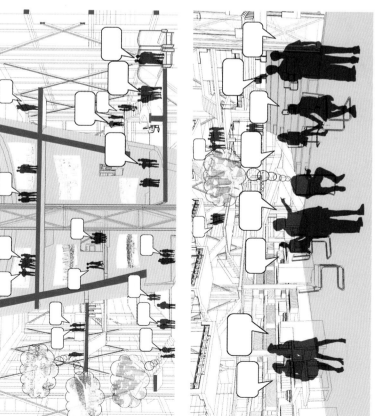
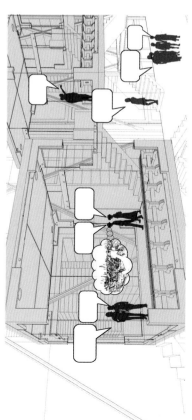

Sectional View

There are several "boxes" adhering inside the museum, these are the room for VR observers, they can easily get the view of every recorded era by the device of VR, and they can compare the past and present by switching from the scene of VR to reality.

校园建筑设计
Campus building category

校园共享广场设计（伦敦大学学院，简称UCL）

伦敦大学学院城市设计 offer 李怡萱

本设计选取校园中的一块空地，将公共空间充分利用，设计成供师生交流休憩的绿地公共景观，同时将建筑功能安排到地下空间，既不占用校园公共空间用地，又将实用的功能巧妙植入公共场所下方。

Communication Playground

Time:Autumn Semester-2014(senior year)
Duration:Three Weeks
Location:Shandong University of Science and Techonology,Qingdao,Shandong,China
Critics:Fei Xia

When graduates from universities look for a job in society,they are supposed to know professional konwledge of some majors.However,in universities,students only concentrate on things about what they are major in,they merely know nothing of others.This situation is bad for students to find what they would like to do as their occupation.

This communication playground is aiming for students studying different professions in universities to exchanges what they know. In this way,they can broaden the knowledge and make new friends.

广场透视图

Analysis of the number of people

There are several people staying in this square. That is to say, this area is useless. So we make three curve graphs to show the propective effect of the number of people:
One is the changes of the number of teachers at the square;
Two is the changes of the number of staff;
Three is the changes of the number of students.

Teachers

Students

Staff

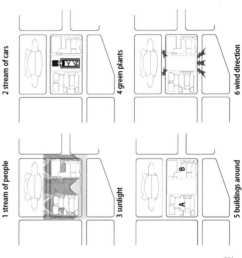

Site Conditions

1 There are four orientation of people. First is from the north of this square. Second is from the south. Third is from two buildings.

2 There are two main roads in north and south of this sqaure.

3 There are enough sunlight all year round because there is no building at thr south of this area.

4 There used to be less green plants at the square, and most floor are made of concret.

5 There are different college in both sides of this area. A is school of humanity and law, B is school of architecture.

6 The wind is joining together in this area and separating at the north of it because of the existing of the building in the north.

1 stream of people

2 stream of cars

3 sunlight

4 green plants

5 buildings around

6 wind direction

Design Conception

Rest Space: There are four forms spaces for people to relax,

1 We make a outside space to give studengts a chance to be close to nature, and enjoy the different scenery of four seasons.

2 We creat a small green space to adding the combination of inside and outside.

3 At the basic of enough plants in this place, we provide a chance for students who are tired of reading books to take a rest by seeing nature beauty indoors.

4 We make some circle components on the ground to ensure there are good sight under the ground.

5 We broke the bounndaries between two spaces by bookcases.

6 We make some wide space and provide some rest place to promote communication between students in different majors.

Human and Nature:
We create three ways to let people to be closed to nature when they go outside and inside. The first one way is enjoying the shadow of trees when it is hot outside, the second way is see the scenery of nature, the third way is to have a relax under these big trees.

1 Outside Space

2 Outside Space

3 Landscape

4 Sunshine

5 Reading Space

6 Chatting Space

Location

The location of base is in Shandong University ofScience and Technology, Huangdao,Qingdao,Shandong,China. It is at the square between two different college of this university, and there is a tall building at the north of this square. There used to be several plants, a small river and two openings in this square.

Shandong

Qingdao

Huangdao

University

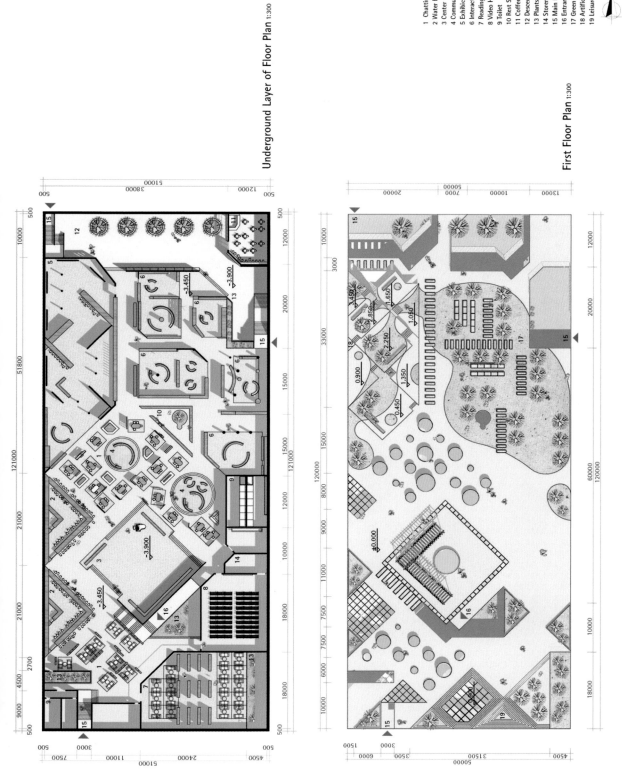

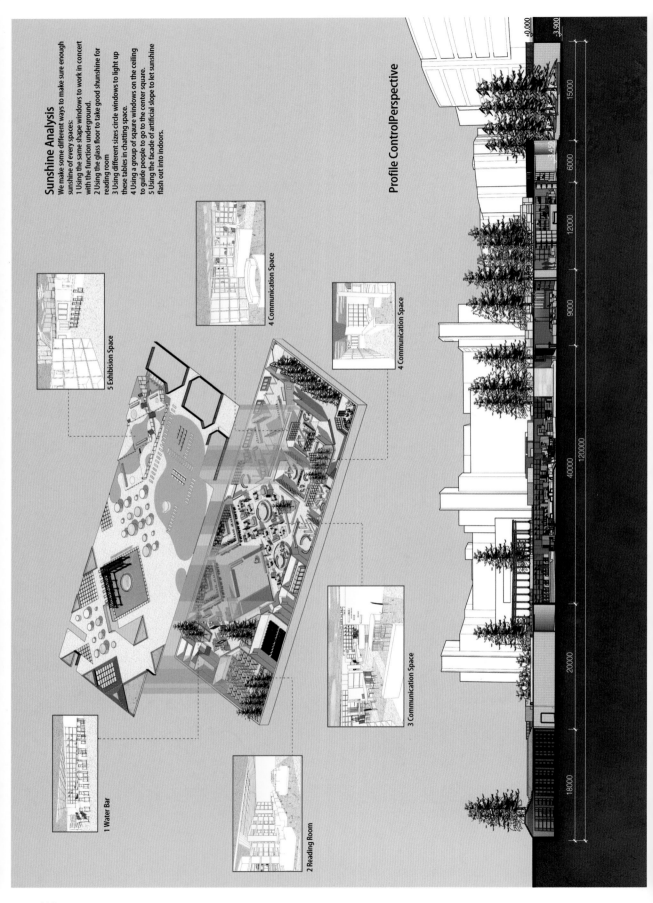

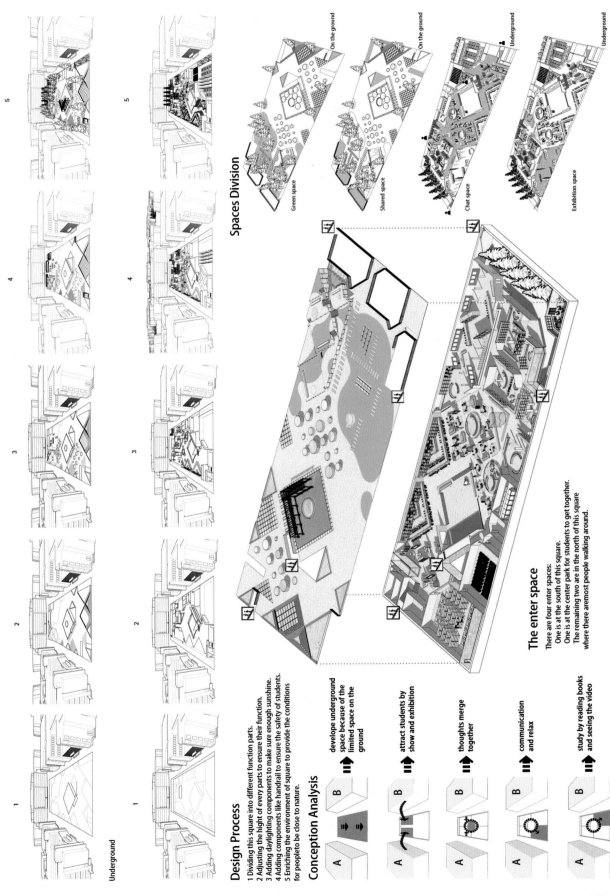

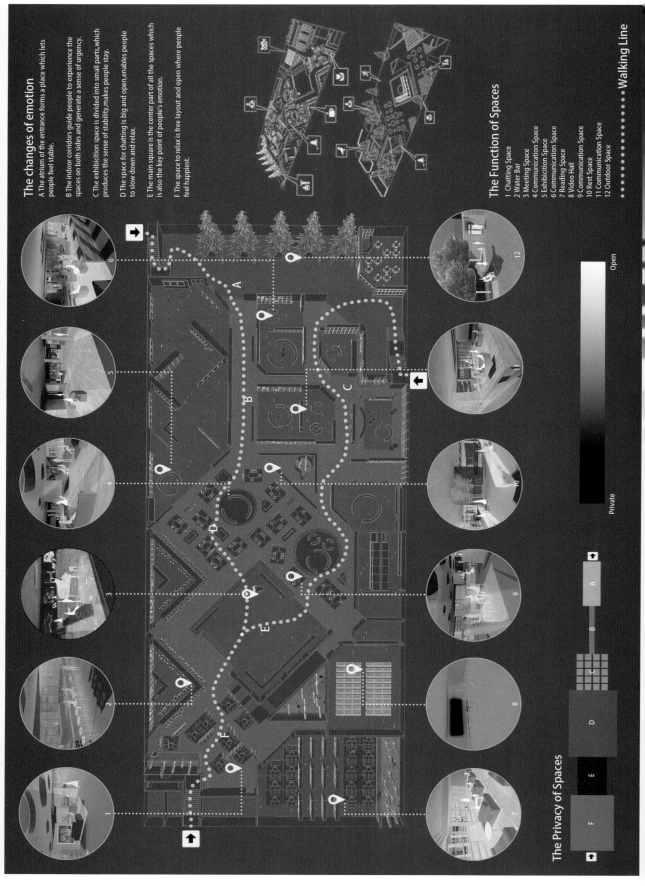

校园港口——燕山大学学生活动中心设计
（米兰理工大学）

米兰理工大学建筑学硕士 许鹏辉

CAMPUS HARBOUR
COLLEGE STUDENTS' ACTIVITY CENTER DESIGN

04

Site area: 11720m²
Time: 8 weeks, 04.2016
Location: Qinhuangdao, Hebei province, China

Design Description: Yanshan University requires a student activity center to meet the students' need of extracurricular activities. It combined with the whole planning and layout of the building of the west campus, which unifies the architectural space with an circle inside. The building wall of the first floor is made of glass material, and the roof window is for lighting, which put the building into the water view and closely linked it up with the center and the surrounding environment. The inner division of the activity center is a large light transmission exhibition space, where there is a continuous landscape square outside to provide students with a place of more possibility to make the activities into occurrence.

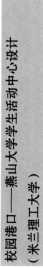

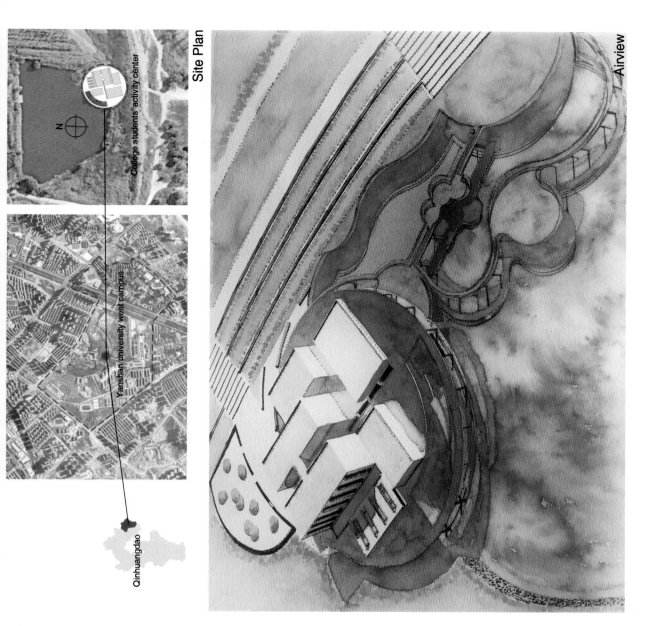

Site Plan

Airview

方案背景介绍及鸟瞰图

Site Analysis

Background

Yanshan university west campus planning and construction of the library and the hill park, base is located in between them and school districts. Schools need to build a university student activity center to contact them, let the whole campus become more dynamic. So what should we do?

Location analysis

- Student apartment
- Study buildings
- Restaurant and Sport building
- Library
- TaShan park

West campus planning is based on a great circle center, buildings in the small circle around, base is located in the center of the circle, the transportation is convenient. Nearby the base is TaShan park and lake, natural condition is superior.

Surrounding environment

- Buildings
- Soil
- Greening
- Lake

Building the surrounding natural environment is superior, in how to use these conditions are considered in the design of the key.

Gemometric Form Generation

1.SITE
Base consists of partial waters and central open space.

2.ENTRANCE
According to the traffic to determine the main entrance and functional layout.

3.GROUNDFLOOR
Building the first floor with a circular glass box to uniform building volume.

4.WATERSCAPE
Building section extends into the water, water gallery has a good field of vision.

5.DAYLIGHTING
Skylight and inner courtyard for construction provides ample light.

6.BRIDGE
The main entrance to the square by the scene of the bridge connection.

7.OPEN SPACE
The sunken square and fountain square constitute the entrance landscape diversity.

8.FINAL SOLUTION
A full of activity possibilities of open university student activity center.

Concept

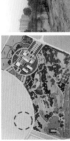

Campus construction meet the planning of the circle, i also try to in the center of the base outside the building outline with circle is unified.

Building first floor exterior wall with glass materials, bring in more light, let the building more transparent.

校园建筑设计
Campus building category

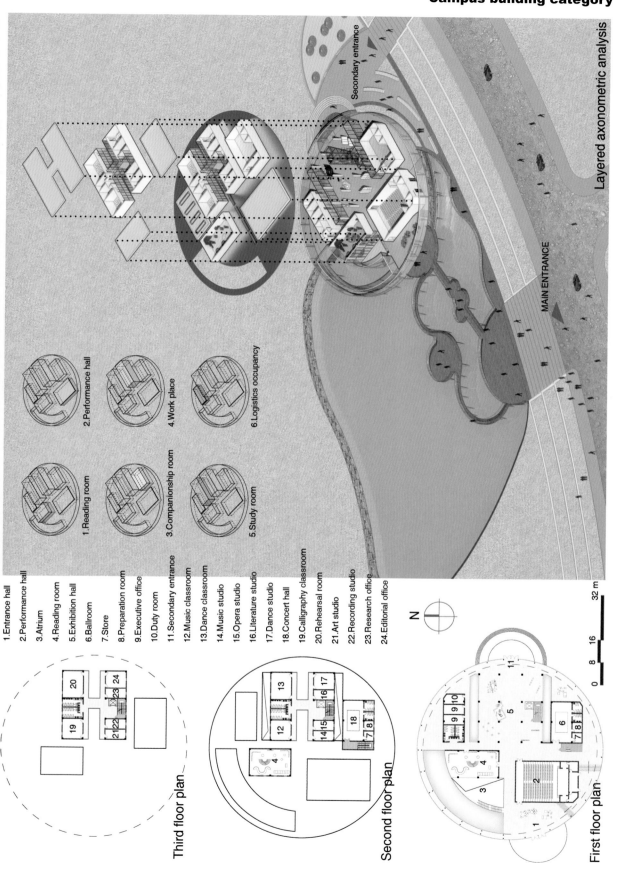

平面展示及功能分析图

Outdoor leisure platform, landscape bridge, roof garden, building glass skin, enables the rchitecture to deliver the messages of "open" and "embracing". Multiple functionalities are fully integrated and can be flexibly transformedfrom one to another including cozy places for leisure and entertainment, indoor activity and display rooms. In the meantime, roof daylighting and the design of the inner courtyard enables indoor get enough light.

1-1 Section and detail

校园建筑设计
Campus building category

Courtyard perspective

Exhibition hall perspective

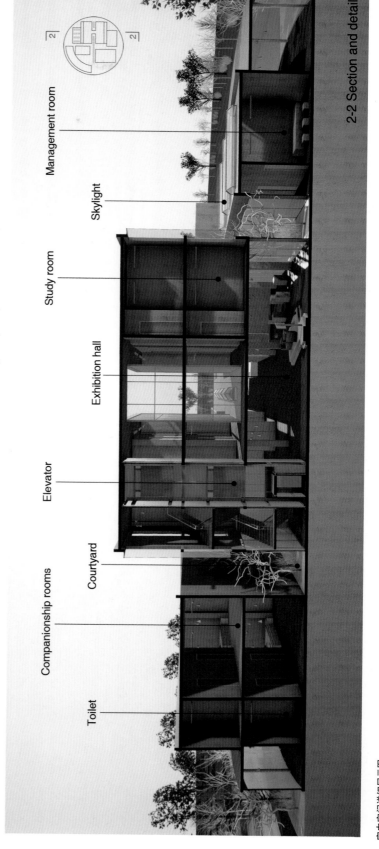

2-2 Section and detail

Management room · Skylight · Study room · Exhibition hall · Elevator · Companionship rooms · Courtyard · Toilet

室内空间详细展示图

Conclusion:

Public building should not be a closed concrete box. It should be jointed the two buildings together rather than be isolated confrontation, which would become the visual center of the campus public space. The building is in an open state, so that the public buildings and campus space would get together to create a free, inclusive and active campus spirit. Activity room and library main building lifting up from the ground, and the venue becomes an open campus park. From the floating box, the space of the building grows up and down as in the organic form of nature. Drift away from the very function of the open system of the main building, there are scenes of a series of wonders, which create a poetic space and fun surprises to encourage users to explore new ways of the possession and use of public cultural construction. The existence of these spaces provides the building with a lively vitality, which will allow the building to stay and to become a real campus harbor.

Building night scene graph

Activities around the building

燕山大学建筑系馆设计（米兰理工大学）
米兰理工大学建筑学硕士 许鹏辉

PAVILION REBIRTH

02 ARCHITECTURE PAVILION TRANSFORMATION

Site area: 1788.5m²
Time: 8 weeks, 11.2015
Location: Qinhuangdao, Hebei province, China

Design Description: The Architecture pavilion in Yanshan University construction was closed, and the teaching facilities fell behind. It couldn't meet the requirements of the study of the architecture of teachers and students any more. The plan is opening the building on the original basis of the frame structure, and activating the area enclosed space to meet the activity needs of the teachers and students through a continuous and changing entrance plaza. Roof garden, atrium lighting, rainwater collection and other energy saving building technology have greatly increased the use life of the building.

Site Plan

背景介绍及鸟瞰效果图

Gemometric Form Generation

1. SITE
The site is consist of the original architecture pavilion and a clearing before the main entrance.

2. ENTRANCE PLAZA
Creating a plaza to open old building and accept all kinds of streamline.

3. ATRIUM
Keep some atrium determine the main entrance and used for ventilation and lighting.

4. SINKING SQUARE
Spatial difference processing to create leisure public space for students and teachers.

5. STUDIO
Provide a studio in order to meet the education needs of architecture.

6. OPEN SPACE
Provide a forum for various possible activities and attract more people to participate.

7. ROOF GREENING
The new pavilion reduce energy consumption to meet the requirements of green building.

8. FINAL SOLUTION
A new green architecture pavilion that keep half the building volume.

Site Analysis

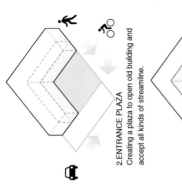

Teacher apartment
Study buildings
Restaurant and Sport building
Student apartment

= = = Teacher route
= = = Student route

Streamline analysis

Base is located in YanShan university east campus teaching area, the south is the student life entertainment area, the west is education study area, and the north is teachers' living area, the east is the sports activity area. The dealership route of the teachers and students of the pedestrian route is completely separated.

Building Analysis

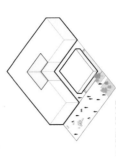

The original architecture pavilion for square concrete frame structure building with the atrium, on both sides for large green space, the shadow of the building have a Boston ivy parasitic all year round, this is the natural landscape, and there is a clearing can be used in front of the building's main entrance.

Advise from school students and my interprets

Exhibition	Cafe	Library	Lecture hall	Ourdoor sports	Indoor sports
Easy to see	Easy to reach	Moderate lighting	Large open place	Adequate sunshine shades	Adequate space
Convenient transportation	Beautiful scenery	Quiet atmosphere		Open fileds	Varieties of sports

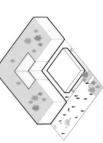

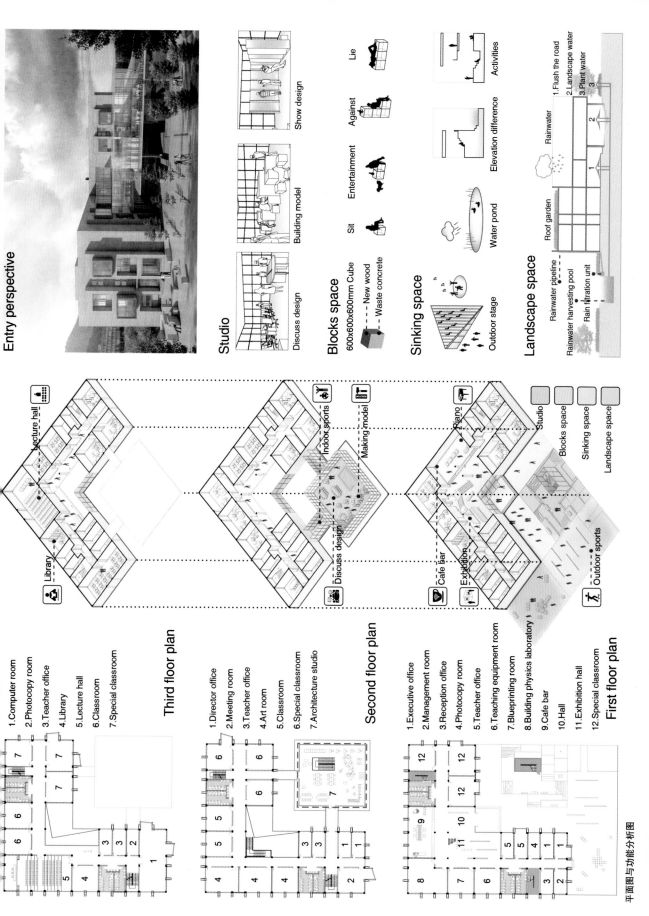

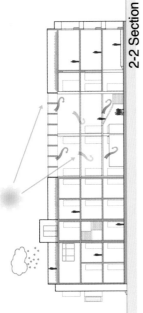

2-2 Section

Outdoor landscape square, atrium entrance, roof garden, open studio, enables the architecture to deliver the messages of "open" and "embracing". Multiple functionalities are fully integrated and can be flexibly transformed from one to another including cozy places for leisure and entertainment, indoor study and display rooms. In the meantime, roof garden, atrium daylighting is ventilated, rainwater collection system maximize reduce building energy consumption.

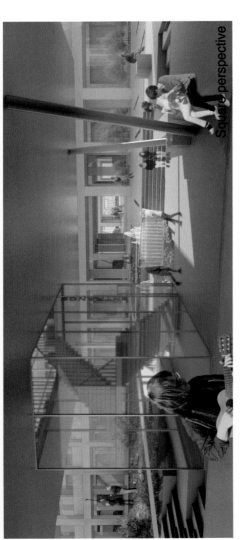

Square perspective

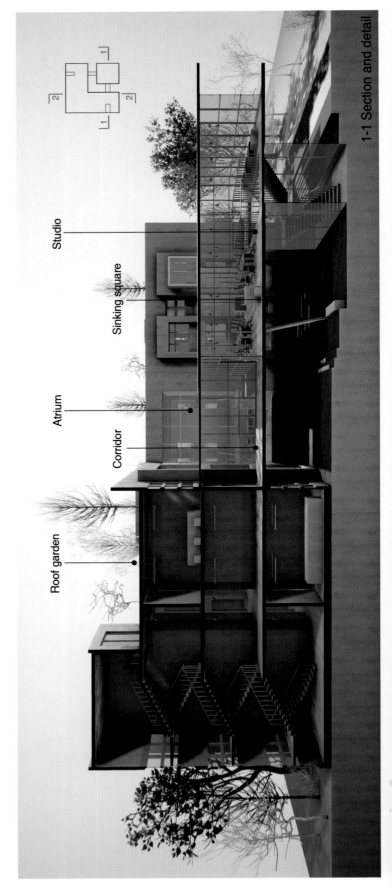

1-1 Section and detail

Studio

Sinking square

Atrium

Corridor

Roof garden

校园建筑设计
Campus building category

Conclusion:

When the demolished and update of city has been an irresistible trend, what I'm thinking about is how to make the process with less cold and short. I hope there is a way to make this process more relaxed, by which we won't lose the precious memories of the story in the process of demolition and reconstruction. I tried to leave the building for a longer time in another way to arouse the common memory of teachers and students of the Department. We don't reject to the quick building of the city, but want to add some human-interest operations into the urban development. It won't take a lot of material, nor it last long. In the design, after the demolition of the building, the concrete wall will be used as smaller concrete blocks and wood combination into the square. The building is no longer belongs to the individual, but a public space, which would not only continue to strengthen the publicity of the architecture department, but also realize a little psychological repair during the broke transformation of the city.

Airview

People view perspective 1

People view perspective 2

The studio

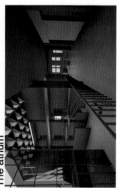

The atrium

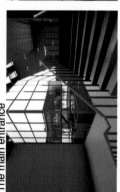

The main entrance

模型透视图

INTERSECTING HUB

FITNESS ROOM ARCHITECTURAL DESIGN

Academic individual work
Workshop in Madrid
Fall 2017

健身馆建筑设计　张晋浩
（英国爱丁堡大学）
（英国伦敦大学学院UCL）
（曼彻斯特大学）

Two lines generates' a point via intersecting with each other, and thus creates a new prototype in architecture. In architecture, it would be really interesting to experiment on the intersection of light, material, sightviewing, pedestrian flows, ect, to see if new prototypes could be created.

As for a small building, subtle changes could be created through using diverse space elements, such as cloumns, railings, walls. The beautiful fitness scenery serves as a stimulus for people to gather and pursue a tranquil mind.

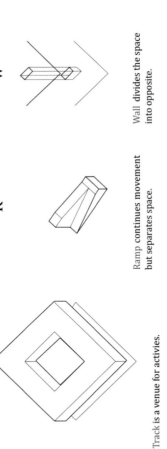

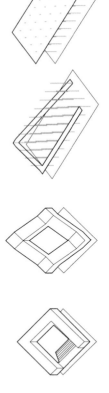

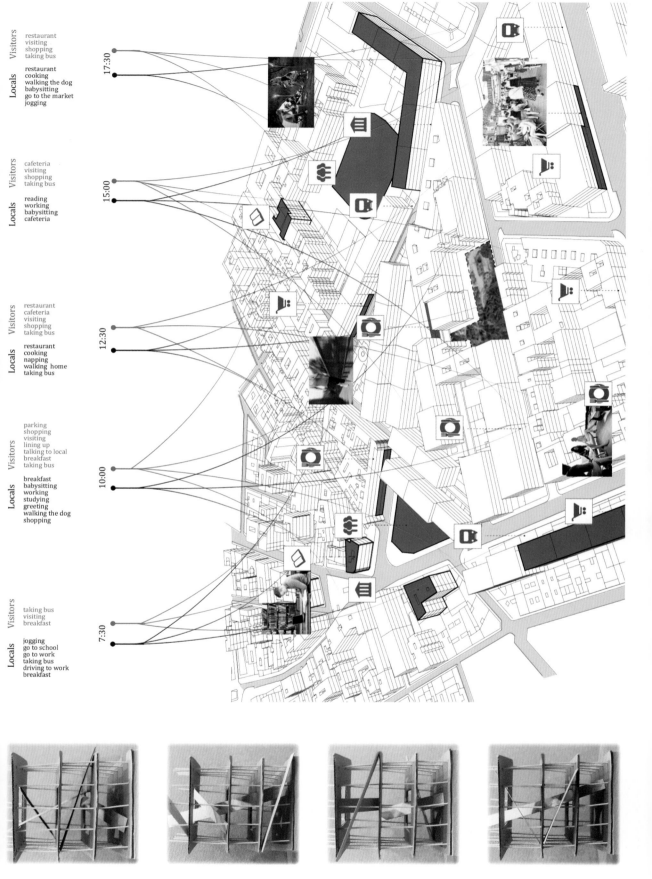

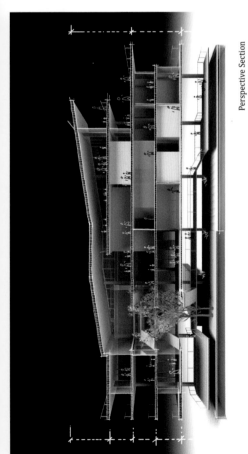

Perspective Section

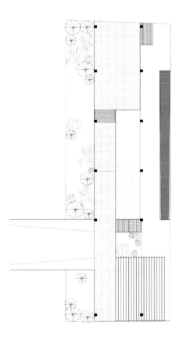

First Floor Plan

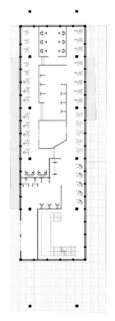

Second Floor Plan

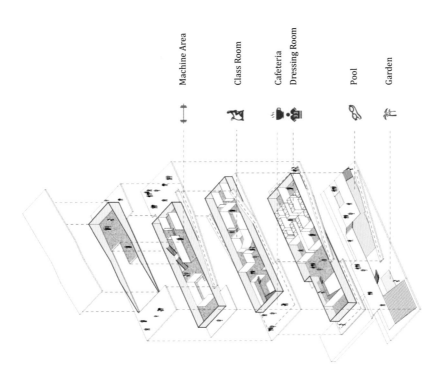

- Machine Area
- Class Room
- Cafeteria
- Dressing Room
- Pool
- Garden

旧建筑改造类设计

台湾猫咪之家——工厂改造设计
（瑞典皇家理工学院）

瑞典皇家理工学院建筑学硕士 吴月

本设计位于台湾的一处已经废弃的旧工厂，当地有优美的自然风光和已经多年不用的工业遗址，并且有很多猫聚集在工厂区域，成为吸引游客的一个新亮点。设计者将工厂改造和新的空间创意结合在一起，不但考虑工厂可能对村民及游客提供新的公共空间的可能性，同时创造性地设计了人与猫交流的空间。为人们更好地了解猫咪，和猫咪共处提供了有趣的空间载体。

Background

Design ideas:

1. Function and the renew of traffic routes of Hou Dong.

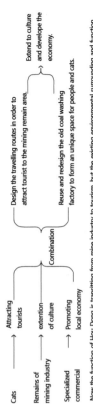

Cats → Attracting tourists → Design the travelling routes in order to attract tourist to the mining remain area.

Remains of mining industry → extention of culture → Combination → Reuse and redesign the old coal washing factory to form an unique space for people and cats.

Specialized commercial → Promoting local economy → Extend to culture and develope the economy.

Now the function of Hou Dong is transitting from mine industry to tourism, but the existing environmental surrounding and function layout are setting limits to the development. The Yi Lan railway seperates village and the mining industry heritage and results into unbalanced commercial development. There should be an organic combination of CAT-MINGING HERITAGE-COMMERICAL DEVELOPMENT supported by new tourism routes.

2. "City of Cats" - The characteristics of tourists' center.

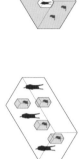

people and cats are seperated in traditoanal exibition spaces.

the cats' territory increased but the interaction between human and cats are still limited.

There should be active and freely dialogue between the two.

The design of tourists' center keeps the framework of the previous coal washing factory while gives it new functions. Cats' territory has been expanded to the east part of railway and modern transparent cooridors and special spaced are used in order to create interesting and various space for tourists and cats. The old builings now have new functions by grafting new spaces.

旧建筑改造类设计
Architecture Renovation category

MASTER PLAN

The mountain town Hou Dong played an important role in Taiwan's mine industry development. How ever this previously prosperous town is declining. The new design is aiming to revive the small town by using tourists' center as a catalysit.

0m 10m 20m 40m   80m

SITE ANALYSIS

The site locates Xin Bei City and adjacents to Hou Dong train station. It has a complex surrounding envirnment. Railway line, city roads and small paths in hilly areas are mixed together. The surrounding district are mainly residential districts with some commercial shops.

TRANSIT ANALYSIS

Railway line

pedestrian route

vehicles' route

local hilly paths.

FUNCTION ANALYSIS

Cat village dewlling
Dwelling with shops
Train
Mine heritage

FUNCTION LAYOUT

PUBLIC FACILITIES

REGIONAL PLANNING

reorganize the previous relative transit chaos function and using the old coal washing factory to build a new tourists' center attracting toursits in cat village. New travelling routes are created to promote the development.

基地分析与场地总平面图

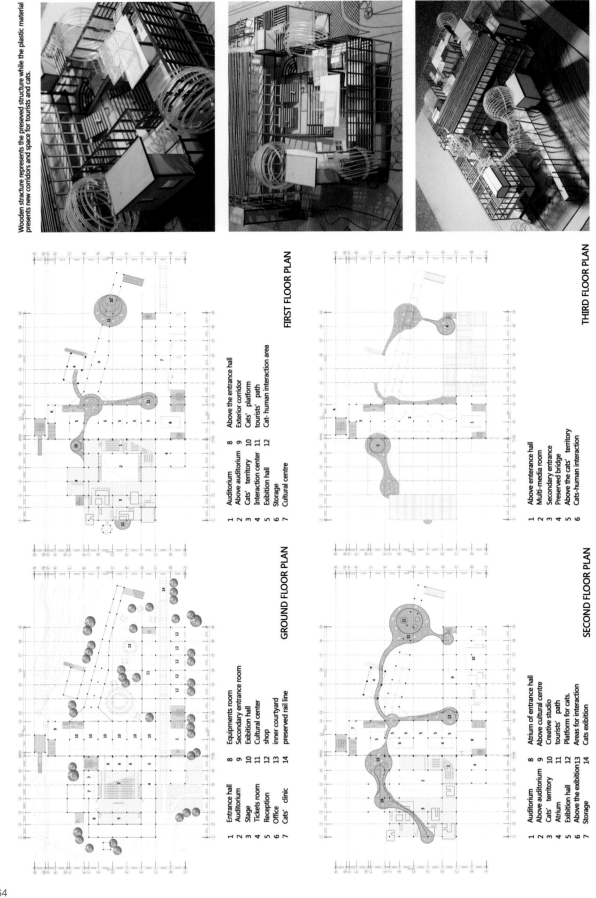

旧建筑改造类设计
Architecture Renovation category

LAYER-ANALYSIS OF CATS' THEATER

"Cats' theater is an important interaction place for tourists and cats. Differed from exibition area and cultural center, lots of platforms and stairs for cats are design in cats' theater. In cats' theater, there are pleasant atmosphere for cat-human interaction. Tourists and cats are both audiance and actors.

PESPECTIVE SECTION PLAN OF CATS' THEATER

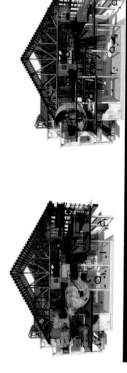

造纸工厂再生——城市天桥展览建筑设计
（香港中文大学）
香港中文大学建筑学硕士　朱启迪

本设计位于香港繁华的Tai Chung路。设计基于一处已经建成的造纸小工厂，将新旧建筑巧妙结合。从城市设计层面和建筑单体层面都提出了解决方案，将工厂风格延续，用新的结构搭建跨越街道的人行步道，将街道两侧的公共空间与建筑联系到一起。城市居民日常的过街行为与参观展览融合在一起。

Finished time: December 2015
During the third year of undergraduate program of Architecture design and
Individual work
Location: Tai Chung Road, Tsuen Wan, Hong Kong

方案简介与鸟瞰透视图

旧建筑改造类设计
Architecture Renovation category

ck terrace & Folding canopy

Site analysis

1. Figure ground

2. Land use
   - Site
   - Residential
   - Open space
   - Educational
   - Factory
   - Footbridge

3. vehicular circulation

4. Public flow

Design theory

In this bridging complex, with abundant supply of paper, the innovative use of paper as building materials can promote local paper making and recycling industry. The design ideals inspire from the natural ventilation. The setback terrace and folding canopy allow more wind into the building. It is really open to public that the circulation is all open or covered. The bridge connects two lots and some activities such as small exhibition and reating area can happen here.

Bubble diagram

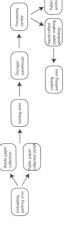

Paper recycling path

Tour path

Concept diagram

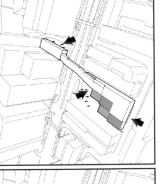
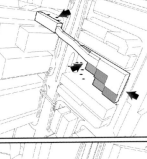

1. The site has two separate lots on two sides of the main road, together with the street in-between.

2. Adding programmes on two sides. The large side which is more public uses the set-back terrace ideal.

3. Making the connection of two sides with the footbridge.

4. The complex has 3 main functions: exhibition, bridge and factory. Covering the network with folding canopy to get more natural ventilation.

5. Adding the landscape element on the ground floor and roof.

旧建筑改造类设计
Architecture Renovation category

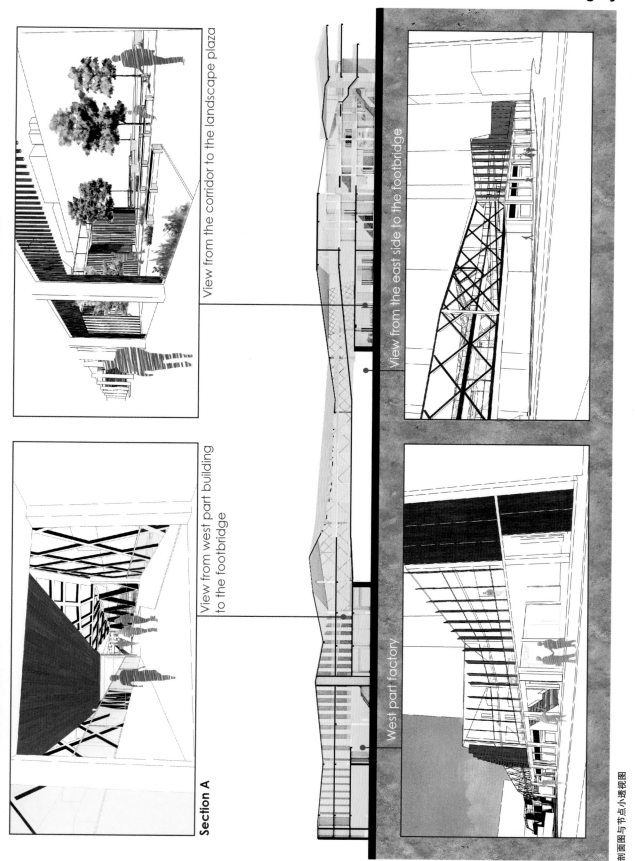

剖面图与节点小透视图

169

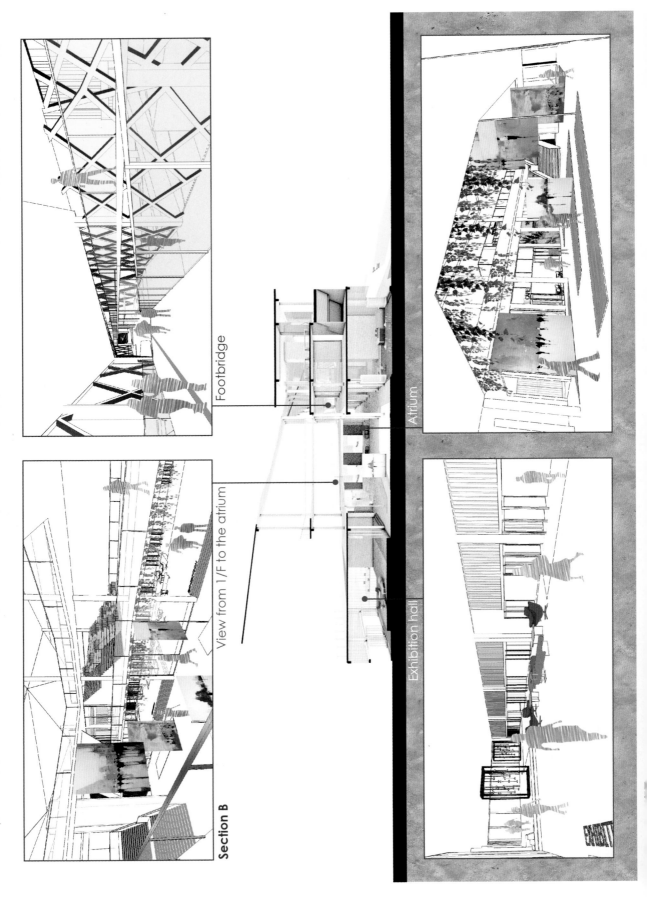

Plans

Main entrance
There is a large landscape plaza at the main entrance which attracts surrounding passengers, workers, residents and students.

View of the whole complex from the second floor deck
The really open building and circulation make people feel more relaxed and interesting. We can see the setback terrace, folding bridge and sloping canopy.

广场透视图与展览馆顶层透视图

Urban design

Period: June, 2016
Location: Harbin, Heilongjiang Province, China
Site area: 35403 m²
Type: Individual Work

哈尔滨铁路局城市设计（米兰理工大学）
米兰理工大学建筑设计及历史史专业 包育鑫

Aerial view

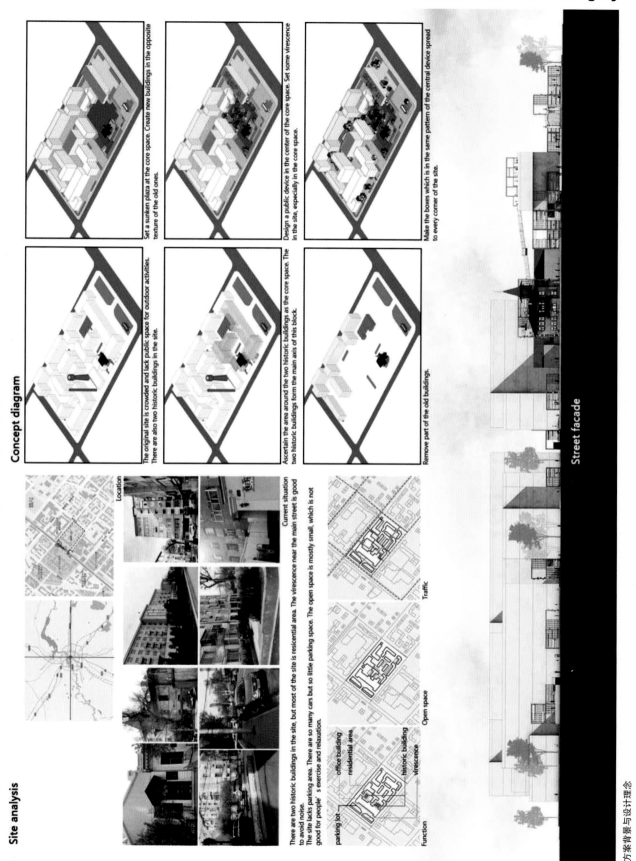

Unit module

The core space in this block is combined by a number of modules which are similar to each other. They are changed from a same basic pattern.

Set a cube with side length of 6 meters. Set main support structure.

Set grid structure for the top surface in order to be stable.

Lay some boards of 1.2 meter-wide, 2.4 meter-length on the top surface. Part of the boards need to be cut properly. The laying area depends on the need.

Set some shelves at two sides of this space.

Set some soil boxes on the shelves in order to plant some flowers.

Set some wooden grid shelves at two sides of this space so that the vine can grow along them.

Perspective View

旧建筑改造类设计
Architecture Renovation category

Inner view of the core space

The plants boxes hung on the shelves in the device can be provided for residents to grow whatever they like.

Different floors are connected with several oblique planes. There are also some interlayers existing.

On the first floor of the device, there is big stairs in the center, also in the atrium. It is next to a big tree.

People can just have a rest, do some reading or waiting for their friends. They can also go through the stairs to the second floor.

There are a wide range of plants in this device in order to make people much closer to the nature. It also provide a place for people to hang out and a great form of relaxation. It aims to change the way of living in nowadays.

Conversion process—from summer mode to winter mode

Part of the oblique planes are hung out of the device. People who stand on it can have a good view of the block.

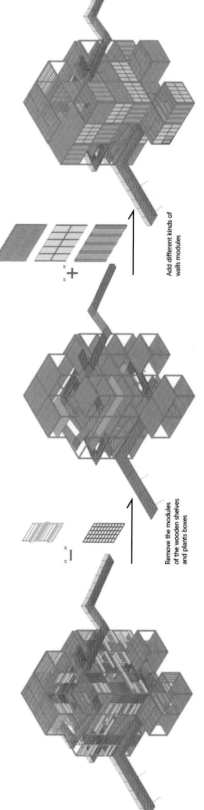

Add different kinds of walls modules

Remove the modules of the wooden shelves and plants boxes

室内透视与冬夏复模式切换

Traffic analysis

Explosive diagram

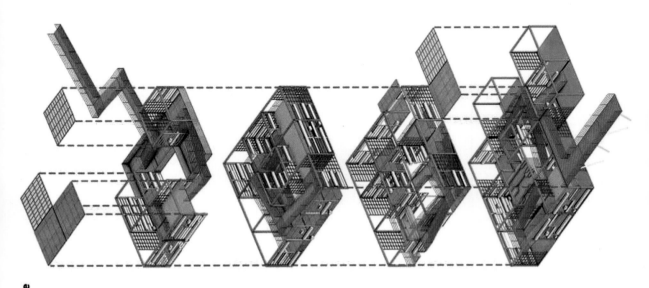

Summer mode

高层类建筑设计
High-Rise Building category

旋转空中花园设计（伦敦大学学院）   吴月
获得伦敦大学学院（UCL）offer

BACKGROUND

Rapid urbanization has promote the development of high rise building, but the high density in urban center and architectural monotonous result in unpleasant city environment for citizen to live. How to creat beautiful high-rise buidling offering harmous living and working places while also demonstrating identity is a key issue for designers.

CONCEPT

Coordinate the geometric changes with eco-principles, so people can enjoy the green scenery in different attitude and promote te vertical greenary development of urban areas.

GEOMETRY GENERATION

1.rectangular site

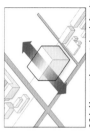
2.Cuboid geometric generated, main facades facing parks.

3. forming tower and Podium.

4.dividing the tower into units.

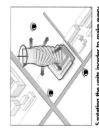
5.rotating the units inorder to make every unit has good view.

6.desiging the front square according to the pedestrian flow.

方案理念演变和模型展示图

MASTER PLAN

The site locates at the city center in Zheng Zhou, Henan Province in China. It is surrounded by governmental offices and commercial district in prosperous city center. Two main roads passing by the site and 2 city parks are on boh south and north side of the site. The design is to create a new high-rise hotel serving as a landmark in this area.

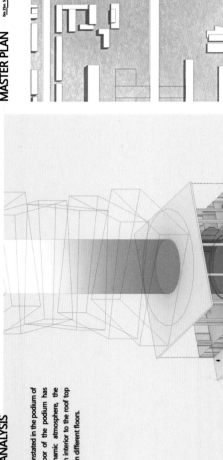

PODIUM LAYER-ANALYSIS

The eco-ideas are also demonstrated in the podium of high-rise building, every floor of the podium has exterior platform with dynamic atmosphere, the greenary has expanded from interior to the roof top and creating green gardens in different floors.

public route
service route

5 F Meeting
4 F Gym +office
3 F Exbibition and commercial center
2 F Dinning hall
1 F Entance hall +cantine

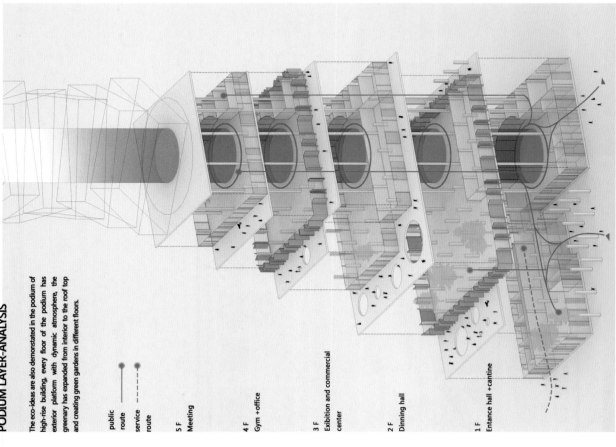

Eleventh Floor Plan

1 Elevator room
2 Cafe
3 Leisure space
4 W.C.
5 Gym
6 SPA center
7 Exterior leisure space

Fifth Floor Plan

1 Elevator room
2 Leisure space
3 Hotel room

Sencond Floor Plan

1 Elevator room
2 Lobby
3 Escalator
4 Stairs +W.C.
5 Service office
6 Exibition
7 Bussiness center
8 Green roof platform

Ground Floor Plan

1 Entrance hall
2 Lobby
3 Elevator room
4 Stairs+W.C.
5 Escalator
6 Service office
7 Meeting room
8 Kitchen
9 Cafe
10 Cantine
11 Parking site
12 Parking site for service
13 Entrance of basement gargage

设计平面图和设计透视图

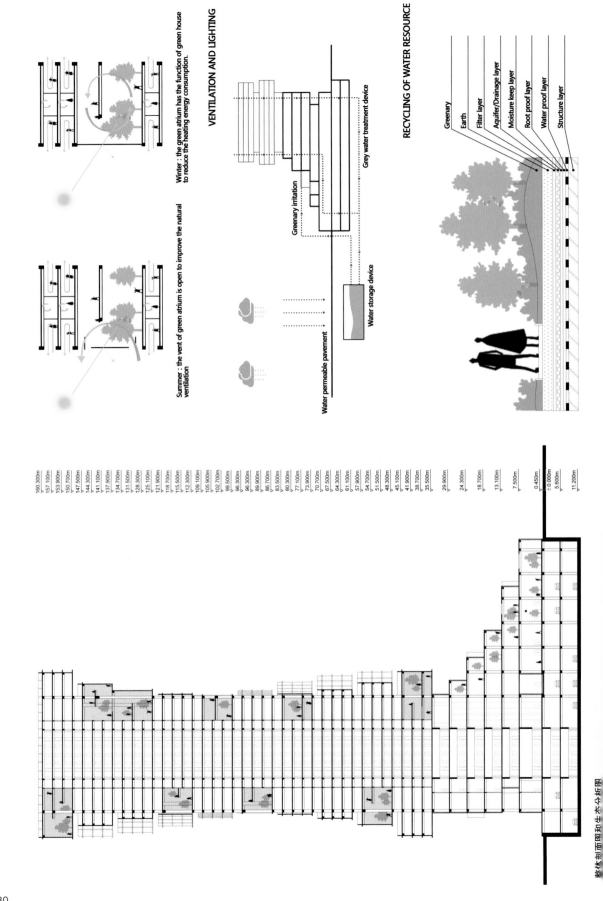

商住综合高层设计（墨尔本大学）

建筑学硕士　刘宇恒

墨尔本大学

The Mixed-use Building

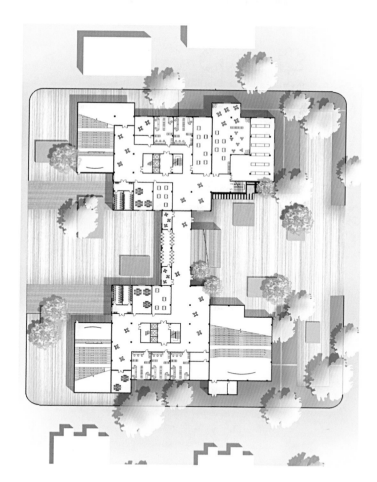

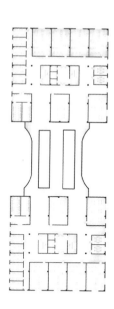

Design idea

Building is not only the building itself that designed on the red lines of lands, and the horizon of an architect should also not limited in the red lines. The more important attribute of building is that it is a part of the city and is the extension of city.

The design starts with the view of campus planning. From the planning, Northeastern University has very structured central axis, and one key in our design is how to echo the axis. The writer designs a set of double traffic system in the axis. The double traffic not only can be for the use of building in the red lines, but also can provide a set of core, public, active traffic system for the whole campus, and bring vigor and vitality to the campus.

In addition, the design not only values the macroscopic concept design, but also researches and designs the indoor scale of comfortable space experience for people. The spaces are more flexible and interesting, and can promote the communication and interaction of teachers and students, and in favor of forming good academic atmosphere.

设计理念和首层平面

The Mixed-use Building

Design Concept

The design starts with the view of campus planning. From the planning, Northeastern University has very structured central axis, and one key in our design is how to echo the axis. The writer designs a set of double traffic system in the axis. The double traffic not only can be for the use of building in the red lines, but also can provide a set of core, public, active traffic system for the whole campus, and bring vigor and vitality to the campus.

Building symmetry echoing the axis symmetric relation of terrain

Considering the concentrating of wind power and the requirements of ventilation of building, hollow the middle to decrease the resistance of wind, and increase the stability of building.

The Axis symmetric relation of terrain

The lighting of the southeast direction is sufficient, and put the rooms that need key lighting in the southeast direction and decrease the energy consumption.

People stream evacuated relationship with surrounding building and sites with the center of square.

Traffic lines status around building and distribution status of inlet and outlet of building.

Building bulk and height limit

Lighting determining the building partition height

Building symmetry echoing the terrain

Hall bulk choosing and connection

Building corridor connection

Complete blocks and flow of stream of people

Small block of building generation

The improving of building and structured site

Office area

Exhibition hall and conference hall

Active region

Art classroom

CONCEPT GENERATING

1. Modern building is superimposed one floor by another. Circulating superposition has no free flavor and also lack for interpersonal communication, making building lack for the atmosphere like courtyards.

Inspiration is from life. Simple drawing cabinet can bring the communication of above and below, and the depth of drawing can drive the interpersonal atmosphere, and also can provide platform for interpersonal communication, to increase the free sense that building should have.

Put the principle of drawer into the building floors. Every floor is a drawer to make the mismatch of every floor, and to make the flexible docking of interpersonal and floors and increase the flexibility and enjoyment of building.

Extend further our ideas, and make building not the unidirectional floor but the three-dimensional transportation, the communication of above floor and below floor, communication of floors of different bulks and communication of mismatched floors, and create free platforms indoors.

Rendering:

The design not only values the macroscopic concept design, but also researches and designs the indoor scale of comfortable space experience for people.

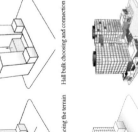

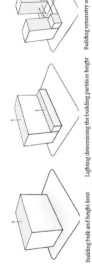

高层类建筑设计
High-Rise Building category

The Mixed-use Building

Design Ideas

The design starts with the view of campus planning. From the planning, Northeastern University has very structured central axis, and one key in our design is how to echo the axis. The writer designs a set of double traffic system in the axis. The double traffic not only can be for the use of building in the red lines, but also can provide a set of core, public, active traffic system for the whole campus, and bring vigor and vitality to the campus.

Rendering: Planting green plants with the carrier of high building can not only beautify buildings, but also make building save energy.

Rending: In the current situation of increasing scanty lands, it is a favorable measure of beautifying natural environment.

模型展示与结构大样图

高层医院设计（墨尔本大学）
墨尔本大学 朱冠宇

HOSPITAL DESIGN
—— Community Central Hospital Design

高层类建筑设计
High-Rise Building category

Period : May.2016-July.2016
Acedemic year : Fourth
Location : Shenyang City, Liaoning
Type : Academic
Tutor : He Sun, Ying Chen

In this project, the type of hospital is a community - based rehabilitation hospital. There are many residential areas and a university around the base, while close to the park. I will rotate some block to let patients have a better vision and let more sunlight enter the building. At the same time, I use the application of landscape architecture and covered roof to create a rich and coherent line.

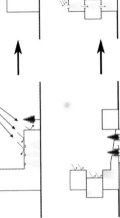

Introduced more sunlight into indoor by changingthe constr -uction form and creating the courtyard.

THE SUNLIGTH ANALYSIS

The base is lo- cated in Sheny- -ang, Liaoning province. There are university, park and reside -ntial area arou -nd the site.

THE CONSTRUCTION SITE

THE REGIONAL ANALYSIS

LANDSCAPE AREA

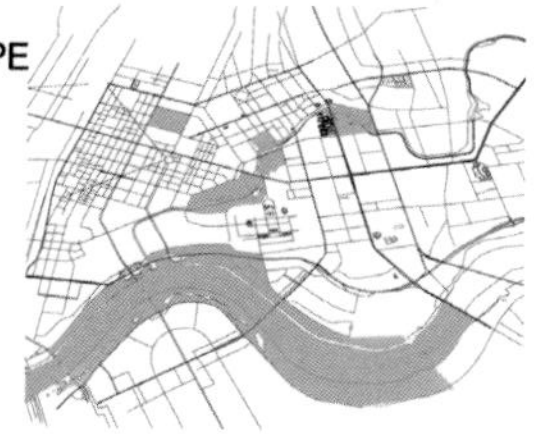

RESIDENTIAL AREA AND HOSPITAL

TRAFFIC FLOW

THE SECTION PLAN

THE LINE ANALYSIS

Office | Activity
Outpatient | Emergency Patient
Inpatients

Architecture Portfolio

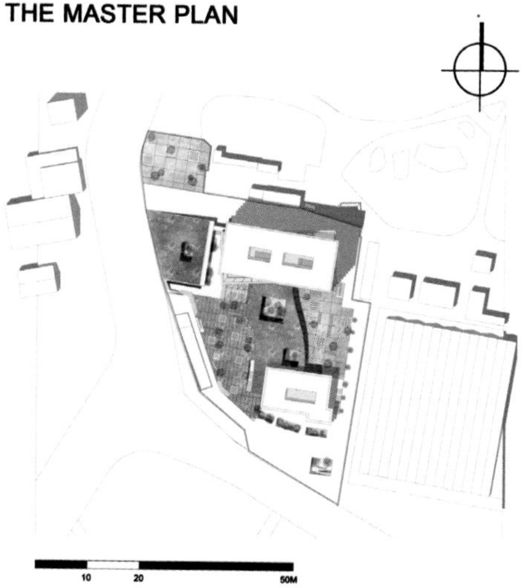

THE SITE ANALYSIS

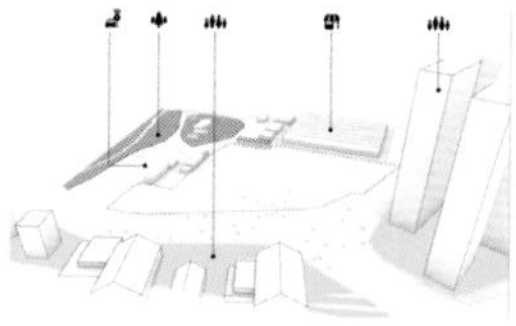

The surrounding area

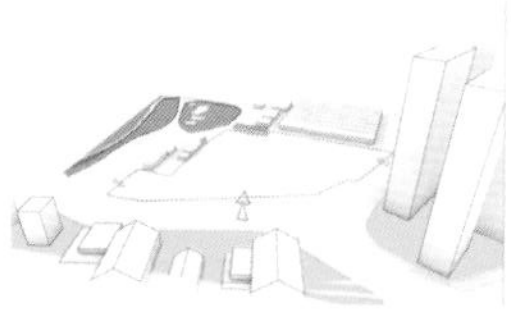

The flow direction

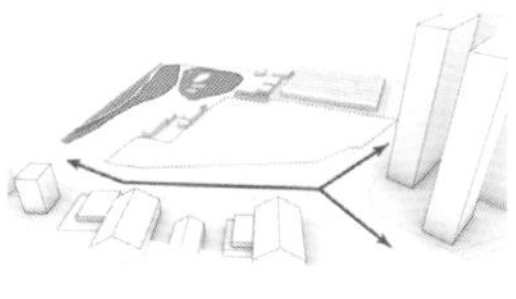

The vehicle flow direction

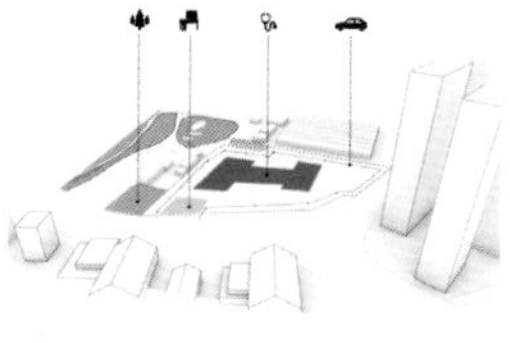

Functional partition

THE MASTER PLAN

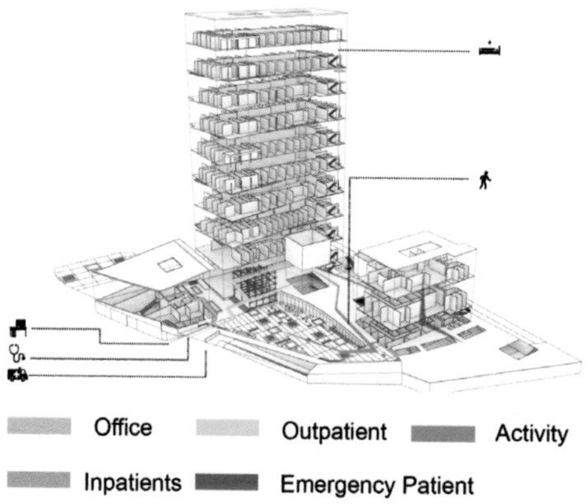

场地分析图、剖面展示图与功能分析图

HOSPITAL DESIGN
— Community Central Hospital Design

高层类建筑设计
High-Rise Building category

Office
Outpatient
Emergency Patient
Activity
Inpatients

THE CONSTRUCTION SITE

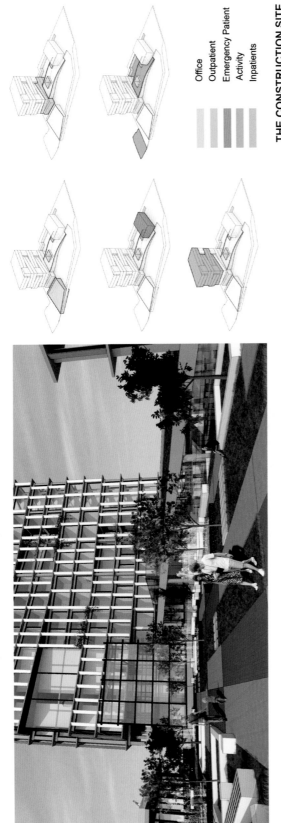

THE CONSTRUCTION SITE

Function partition	Ground subsidence	Block generated	Change block
Block change	Block rotation	Insert atrium	Final

Architecture Portfolio

THE AERIAL VIEW

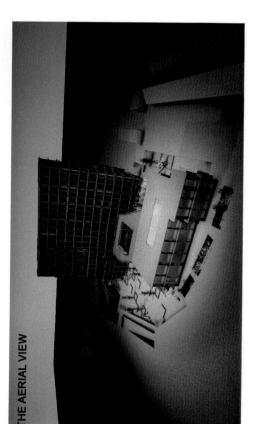

THE GROUND FLOOR PLAN

1. Entrance Hall
2. Registration Office
3. Office
4. Registered Hall
5. Pharmacy
6. Treatment Room
7. Surgical Department
8. Infectious disease Department
9. Gynecology Department
10. Emergency Department
11. Theatre
12. Ward
13. Boardroom

高层类建筑设计
High-Rise Building category

CAPSULE RESIDENTIAL TOWER - The rising landscape

高层胶囊公寓设计（曼彻斯特大学）
获得曼彻斯特大学建筑学 offer 朱启迪

Finished time: June 2016
During the third year of undergraduate program of Architecture design and theory
Individual work
Location: Tai Chung Road, Tsuen Wan, Hong Kong(adjacent to and connect to the bridging complex project)

Design theory

The main programme of this capsule residential tower is a hostel for single travellers or local residents who only need a small sleeping space with low degree of privacy. In contrast, they love to meet new friends in the common or public spaces of the hostel. I explore organization of capsule living units as well as integration of hostel and public landscape in a vertical tower. To attract public use of the upper floors including the rooftop cafes and viewing decks, and to enhance interaction between different users, the tower is designed with fluidity of circulations and visual linkage between different floors.

1. The site is a 16.5m×85m rectangle near the previous complex project.

2. Building the block for about 80m height.

3. Lifting up the top 70m's block and inserting the landscape to create a new 'ground floor level'.

4. Dividing the blocks from function in both vertical and horizontal directions and creating more open spaces on upper floors.

5. Rotating the blocks to get the best view, natural ventilation and daylight.

6. Adding the bridges between two sides for connection and means of escape.

Master layout plan

方案生成展示及场地总平面图

高层类建筑设计
High-Rise Building category

Small capsule
The small capsule which only has one bed can accommodates one person. The size of the small capsule is 2800mm(L)x1500mm(W)x1400mm(H). The capsules are arranged at two long sides. The middle part is the common space such as the open kitchen which is showing in the rendering.

Exhibition hall
The exhibition hall serves for public. The double-layer design creates rich spatial forms for meeting the recreational needs from different visitors.

Park on the second floor
This park defines a new level for public. It connects to the existing footbridge and the bridging complex.

Reading area
Reading area is in the bridge between the exhibition halls.

Plans

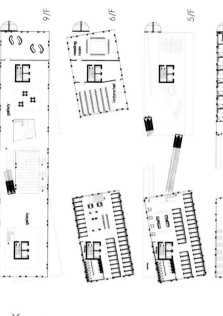

Outdoor lecture theatre
The outdoor lecture theatre is on the tenth floor. Here can have open air concerts, speechers during the day and play football match video at night.

Top level's viewing deck
The viewing decks are provided on top floors nearby the coffee shop. People can have a bird view of the whole tsuen wan district.

Roof garden
People can take the sightseeing lift to the roof garden directly. It also serves as the means of escape by helicopter.

平面图和功能展示图

191

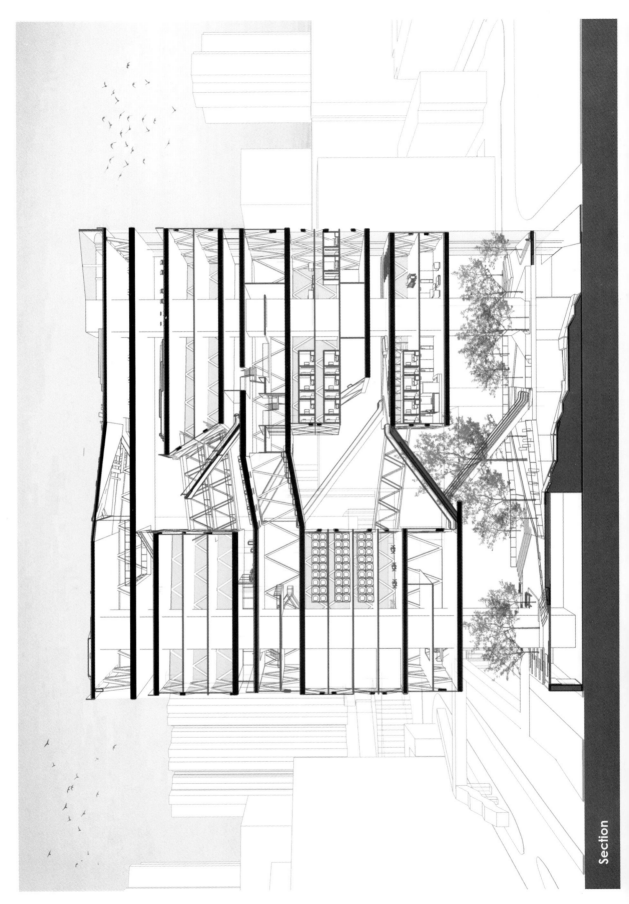

Section

秦皇岛东北大学校园高层设计（米兰理工大学）
米兰理工大学建筑设计及历史硕士 蔡笑革

NEW CAMPUS OF NORTHEASTERN UNIVERSITY AT QINHUANGDAO

Stadio course assignment
4th Year Individual Work
Time Period: 8 Weeks
Location: Qinhuangdao, Hebei Province, China
Site area: 70818 ㎡

方案简介及鸟瞰透视图

高层类建筑设计
High-Rise Building category

Site Analysis

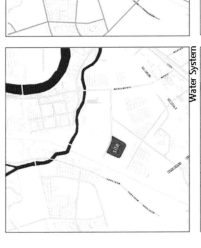

Vehicle Network

Function Distribution

Water System

Surrounding Red Buildings

Surrounding Environment

There are a number of retails and residential areas surrounding the site, which can meet the requirem
-ents of the students and teachers.
Plenty of surrounding constructions are red, which has fromed a style of this area.

Overview

The size of Northeastern University at Qinhuangdao is not very large, it cove-rs an area of 470,000 m² and has more than 800 teachers and staffs.
Now, the small size has restricted the the number of pupils, which hinders the development of the university. The authority decide to extend the size of the university.

Site Location

The new site locates in the southwest to the main campus. There is just one road seperating the these two, so it is very convenient for pupils and teacher-s to study and work.

Analysis Of Main Campus

Water System

Main Constructions

Plazas

Landscape

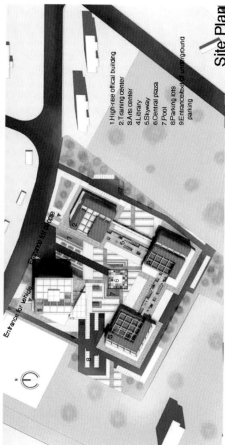

Site Plan

1. High-rise official building
2. Training center
3. Arts center
4. Library
5. Skyway
6. Central plaza
7. Pool
8. Parking lots
9. Entrance/exit of underground parking

Landscape

Public Space

Water System

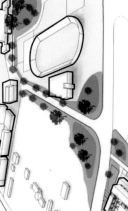

— Circulation of teachers
— Circulation of pupils
— vehicle circulation

Axis of Landscape

Site History
This is an unbuilt site of the branch of Dongbei University in Qinhuangdao. The picture shows the relationship of the new site and the old one.

Rise
The volumn rises according to boundaries and take full use of the available space.

Seperation
Divide the volumn into 4 parts, an official building for teachers and staffs, three teaching buildings for students, respectively.

Revolve
Revolve the official building to be oriented to south, the remainder three buildings are oriented to the direction of the street. These two axis are oriented to the surrounding city texture.

Landscape
The axis of landscape starts from the "Chensi Plaza", extend along with the "Xi Road" and finally arrives at the destination (the central plaza of the new campus.) the landscape does make the relationship between the old and the new be more close.

Connection
The 4 volumns were all isolated originally, but the 3 buildings used by students should be taken full advantage of if connected by skyways. The skyways also planed by viresence, so the students in the 3 buildings can touch the viresence easily.

Style and colour
This area has a distinct feature: a certain sum of the buildings are painted red, there are also some red constructions in the old campus, so the main colour of the new is red too.

Final Solution
The 4 buildings seem to be integration through the axis, colour, texture and the skyways, also, the landscape will be found everywhere.

场地设计及方案形体生成演化图

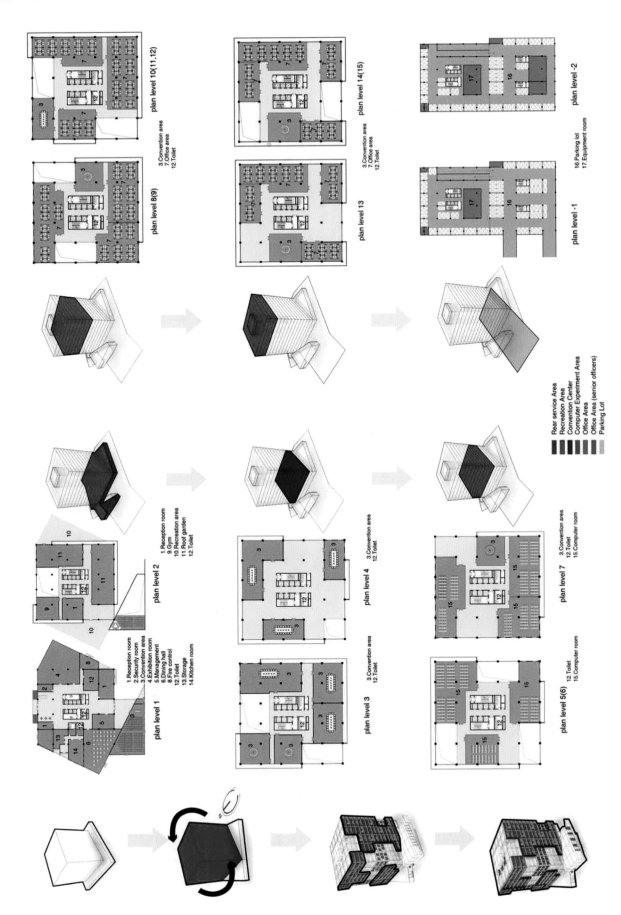

Concept Of Skyway

1. No connection?
2. Use skyway
3. Too dark here?
4. Hole on the skyway
5. No landscape?
6. Plant trees on it

公共步道系统分析图

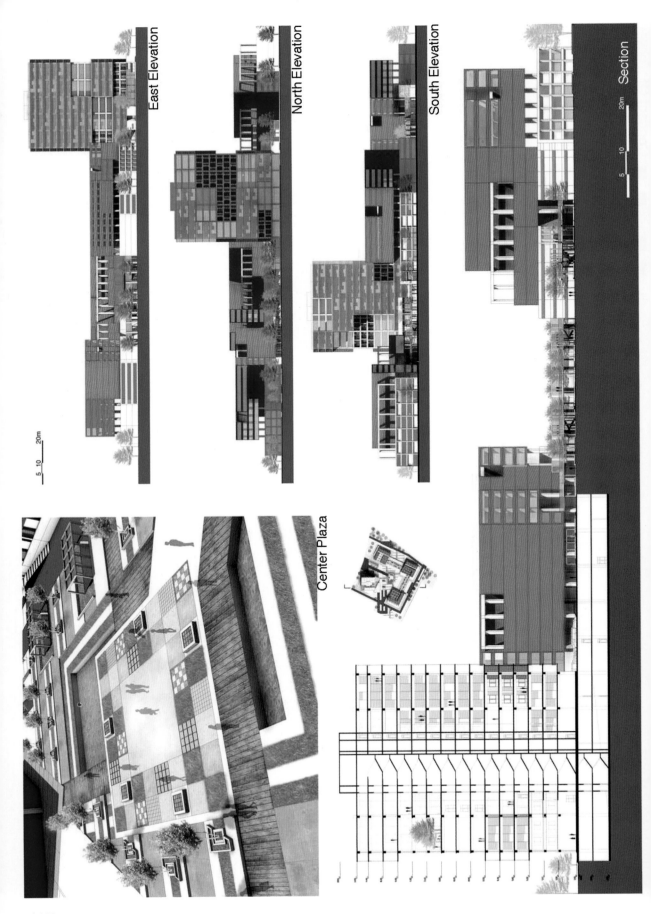

二、城市规划类专业方案

滨水空间城市设计

松花江滨水空间可持续再生设计（瑞典皇家理工学院）
瑞典皇家理工学院建筑学硕士　吴月

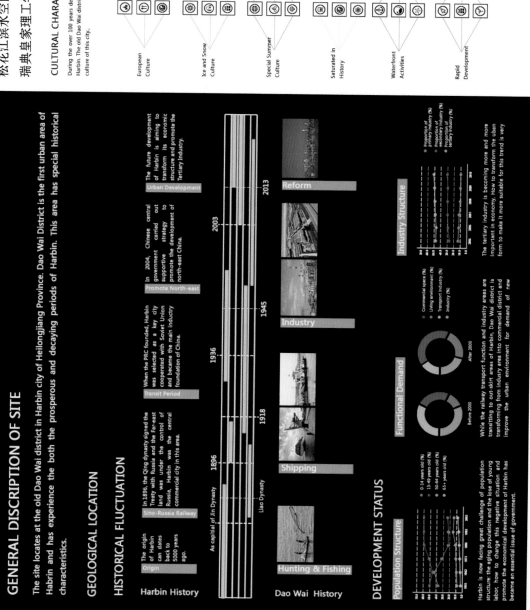

背景调研分析图

GENERAL STRUCTURE

EVOLUTION OF DESIGN

How could we make this area an ecological and harmonious area.

Starting from the existing urban blocks and the environments

Creating three linear public space alongside the three axises.

Generating three landscape axises from the waterfront into the inner urban area.

Generating functional layout

Forming the public transit system in this urban area.

FORMING OF WATERFRONT

Traditional parallel waterfront

Changing the outline of waterfront

Various waterfront shapes attract citizens to this place and create more opportunities for activities.

Integrating the water and waterfront area together

STRATEGIES

1 BEAUTY OF RIVER

Water is an essential natural element, in order to make Dao Wai district a more dynamic and culture-flourished urban area for citizens, the water front area should play an important role of urban revival.

Waterfront Theme Activities

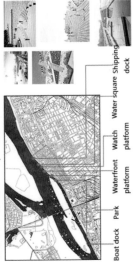

Water playing, Dragon boats | Beer festival | Snow festival | Sino-China cultural activities

Waterfront Theme Landscape Reform

Boat dock | Park | Waterfront platform | Watch platform | Water square | Shipping dock

2 DYNAMIC LIVING

The waterfront reform follows the principles of land mixed-use, high desity use, and ecological reform stratigies. All parts of this area are integrated together as an organic whole.

3 ECOLOGY

Enhancing the ecological environment of the waterfront, including the clean water, clean air, green parks etc. Aiming to create a pleasant waterfront area for the citizens.

BEFORE → **AFTER**

PREVIOUS URBAN FORM

The previous urban district is too intense and lack of green space. The environment is very chaos.

WATERFRONT COMMERCIAL

Integrate urban green and urban blue into this district.

MODERN HOUSING

Integrate the housing area into the urban green and commercial areas.

ENTREPRENEUR AREA

Small blocks and green streets are cooperated in order to enhance the vitality of this area.

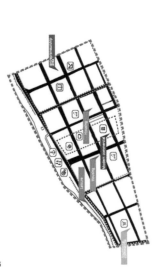

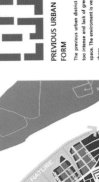

The renaissance of water front of Dao Wai district of Harbin has use the strategie of "river character, ecology, vital living" to integrate the traditional commerical form in Dao Wai district and modern commercial form together. The new creative Entrepreneurship studio and other functions are mixed together in this waterfront area in order to stimulate the urban vitality.

REGIONAL SYSTEM

The waterfront green system and public service system together form the regional system's frame, which has three sub-system with individual axis, the three sub-systems are: "Dao Wai commercial district", "mix-used district of modern residential and commercial" and the "entrepreneurship zone", forming a system with great vitality.

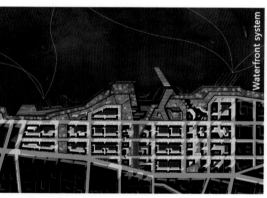
Function zoning

Waterfront system

Green system

Public system

FUNCTION ZONING

The zoning is based on the characteristic of Dao Wai district. The district is divided into three parts:" tradition Dao Wai commercial district", "mix-used district of modern residential and commercial", "entrepreneurship zone", each has its own characteristic.

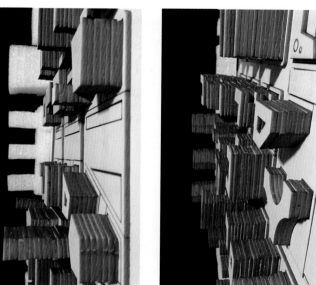

"Tradition Dao Wai commercial district"

urban fabric follows the traditional courtyard form in Dao Wai district, and the commercial courtyard are connected with open urban green system.

"Mix-used district of modern residential and commercial"

This district is formed by central commercial-landscape street and alongside residential area, wich integrated the modern commercial characteristic and dwelling together.

"Entrepreneurship zone"

This district takes advantages of waterfront and aims to create dynamic and supportive atmosphere for the entrepreneurship.

设计特点展示及模型局部透视图

WATERFRONT ACTIVITIES

JAN	Harbin music festival
FEB	Dao Wai cultural exhibition
MAR	Melting festival
APR	Yacht club
MAY	Dragon boat racing
JUNE	Cultural carnival
JULY	Dao Wai traditional food festival
AUG	Beer festival
SEPT	Dock festival
OCT	Commercial exhibition
NOV	Sino-Russia cultural exchange
DEC	Ice-snow festival

STRUCTURE OF WATERFRONT

Shipping dock
Festival square
Citizen square
Historical dock

The waterfront space has a radical effect to the inner urban district, which has three blocks cooridnating with the waterfront

Public Space System Analysis

FORM OF WATERFRONT

- NICE WALK! — Cantilevered Platform — FISH
- IT'S SO CLOSE! — Leisure Center — WATER
- NICE VIEW — Gentle Slope with Natural Scenery
- LISTENING — Watch Platform
- REFRESHING — Boat Dock

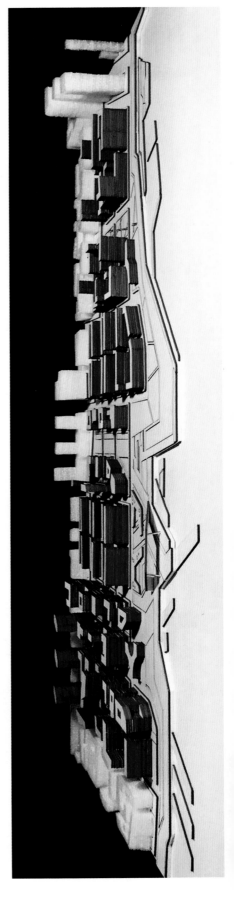

204

历史街区城市再生设计
Historical Urban Area Revitalization

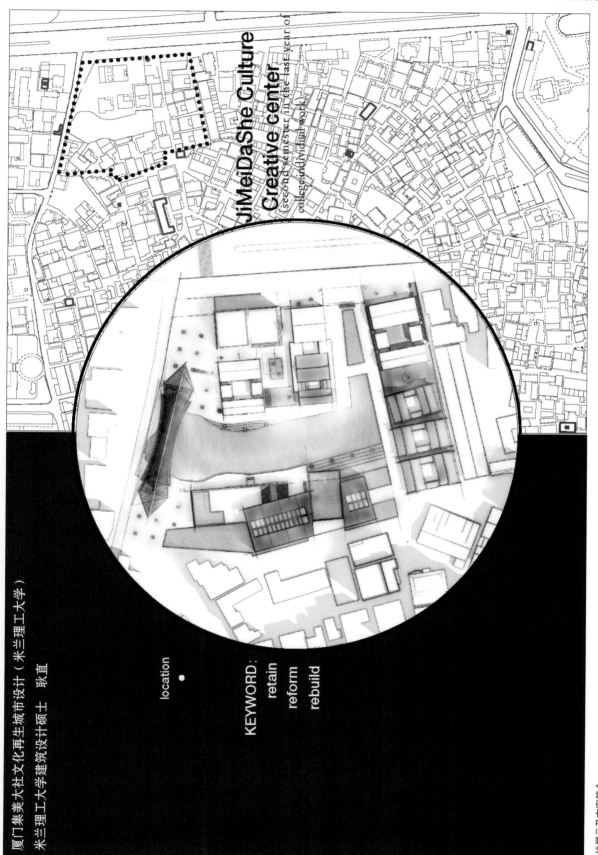

厦门集美大社文化再生城市设计（米兰理工大学）
米兰理工大学建筑设计硕士 耿直

JiMeiDaShe Culture Creative center
(second semester in the last year of college,individual work)

location

KEYWORD:
retain
reform
rebuild

基地展示及方案简介

Bird View 1

Bird View 2

JiMeiDaShe

JiMei means "collect the beauties" in Chinese. DaShe means "an enormous community" JiMeiDaShe has existed for more 100 years.

But now,It has became a terrible problems for inhabitants who is living there because of the mess and bad conditions So I think change the fonction of this area might be useful, to revive JiMeiDaShe

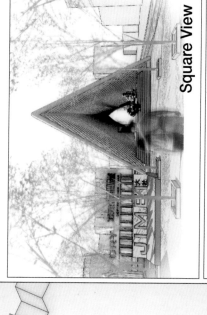

Square View

yard view

Idea Factory (bookstore)

Bridge Gallery (temporary exhibition)

重要建筑的节点图

历史街区城市再生设计
Historical Urban Area Revitalization

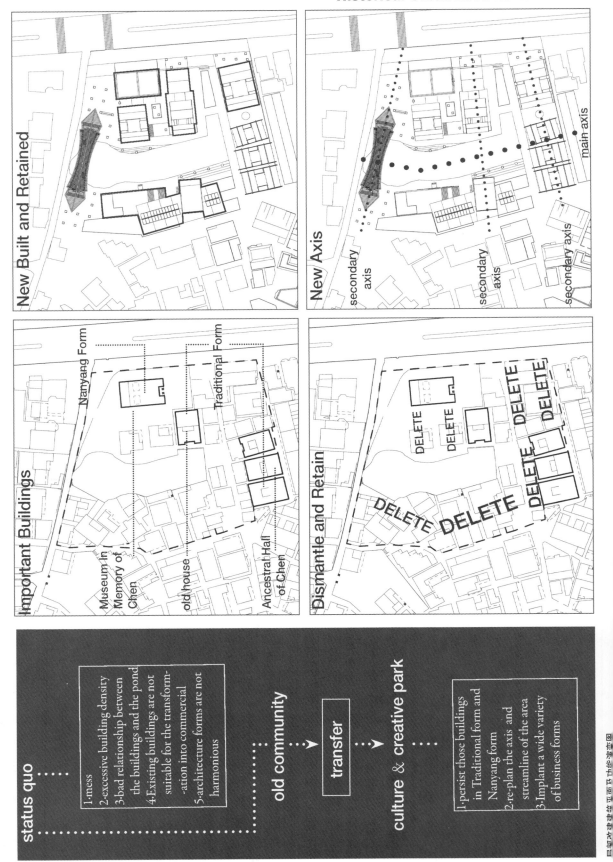

Buildings and The Pond

The pond has been a great problems because of the bad relationship between pond and buildings.Its value has been wasted.
In my plan , this pond is the core of this area.I organiznend the streamline around the water . People can easily walk down stairs and feed fish.
The hight of buildings gradually decrease from road to lake in order to get a open field of vision.

The New Built and The Pond

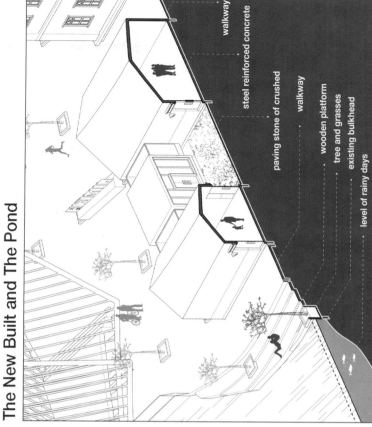

Reconstructive Structure and The Pond

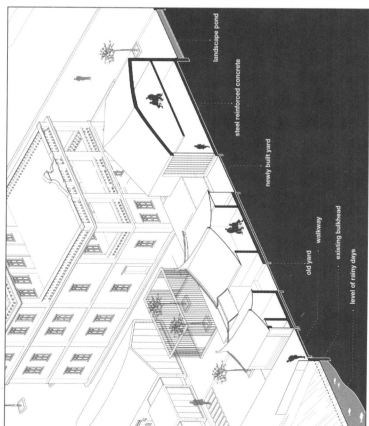

局部改建剖透图展示

View of Idea Factory

idea factory
1. main entrance
2. stackroom
3. warehouse
4. bathroom
5. garden
6. office
7. stairs to the roof
8. cashier
9. ladder glassbox
10. secondary entrance
11. roof garden
12. coffee

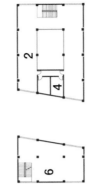

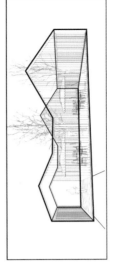

Pick up the outline of traditional buildings and transfer into morden architceture language.
Take a movable wall as main entrance.

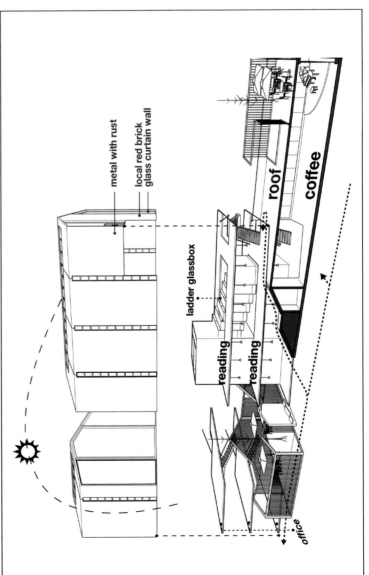

- metal with rust
- local red brick
- glass curtain wall
- ladder glassbox
- reading
- reading
- roof
- coffee
- office

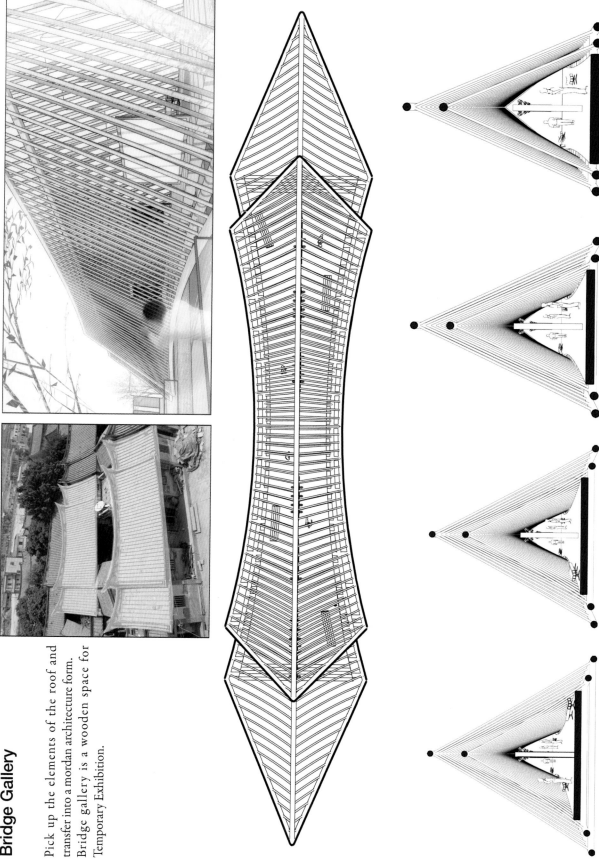

Bridge Gallery

Pick up the elements of the roof and transfer into a mordan architecture form. Bridge gallery is a wooden space for Temporary Exhibition.

历史街区城市再生设计
Historical Urban Area Revitalization

Rebuild and Reform

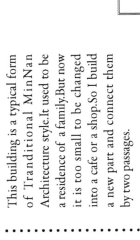

This building is a typical form of Tranditional MinNan Architecture style.It used to be a residence of a family.But now it is too small to be changed into a cafe or a shop.So I build a new part and connect them by two passages.

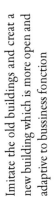

Imitate the old buildings and creat a new building which is more open and adaptive to bussiness fonction

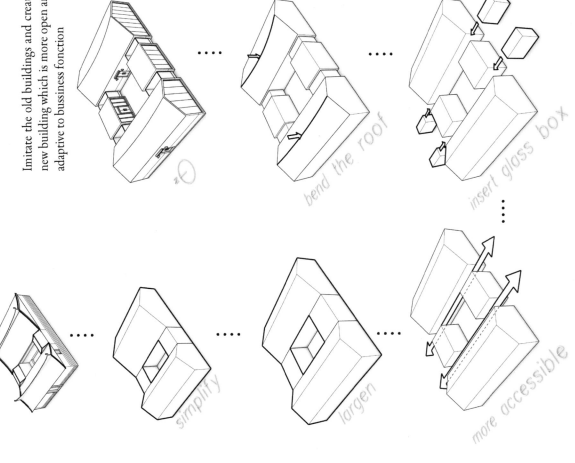

bend the roof · insert glass box · simplify · largen · more accessible

212

URBAN DESIGN
—Renovation Project of PuKou Railway Station

Period : September 2015 - November 2015 (8th Semester)
Location : Nanjing, Jiangsu Province
Type : Academic Work
Tutor : Xi Jianglin

南京火车站改造城市设计（伦敦大学学院）
UCL 建筑设计硕士 苏安琪

PERSPECTIVE RENDERING

Design Ideas: Urban Catalyst

In this semester(9th Semester of Bachelor program), I did a urban renovation project in Nanjing city. The task is to reuse an old unused train station. In order to attract citizens and make it become a new catalyst of this aera, I want to change the train station together with the whole site into a new urban public space. In this way, by integrating this site into the urban transition of Nanjing city,it can bring vitality and cultural diversity back to this area.

River & Lake
School
Factory
Resident
Green Space
Main Road
Highway
Railway
Land Boundry

Site location

 Map of Nanjing in 1988

 Map of Nanjing now

 Map of Nanjing in 1912

 Map of Nanjing in 2004

Since the PuKou Nanjing railway station was built in 1912, it has attracted more and more residents to move to this site.

In 1988, due to the opening of the Nanjing Yangtze River Bridge, Pukou railway station, which has been reused and re-closed for several times , was completely abandoned in 2004.

As part of the cultural protection units,Pukou railway station now only opens to photography crew.

This project, aiming architectural adaptive reuse, is to transform and utilize the space of the old station. And the project also has new premises that it should have convenient transportation to meet the requirements of the function.

背景分析与设计鸟瞰图

历史街区城市再生设计
Historical Urban Area Revitalization

URBAN DESIGN
RENOVATION PROJECT OF PUKOU RAILWAY STATION

BEFORE

AFTER

- recreation square
- exhibition area
- commercial area
- commercial area
- commercial area

RENOVATION PROJECT ABOUT PUKOU RAILWAY STATION

KEY PLAN

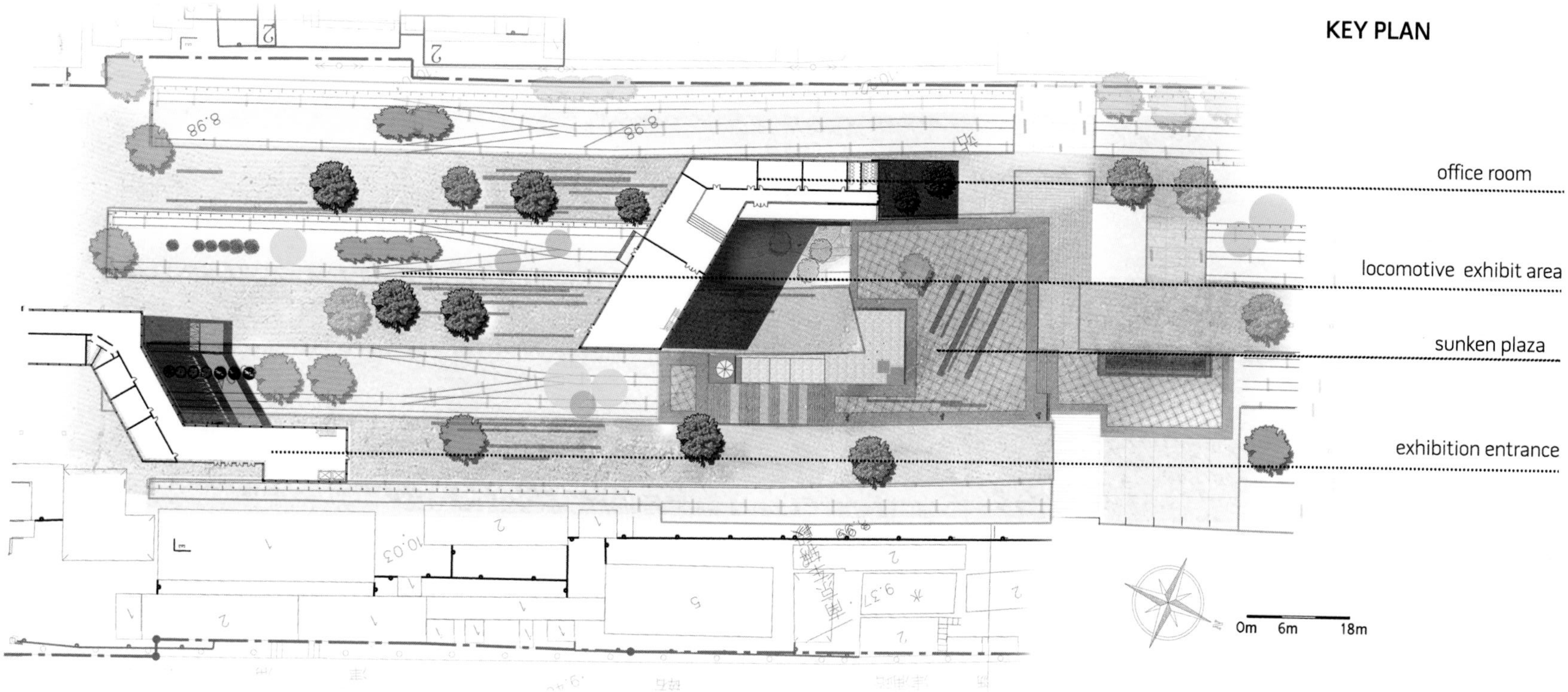

- office room
- locomotive exhibit area
- sunken plaza
- exhibition entrance

DESIGN IDEA AND DEVELOPMENT

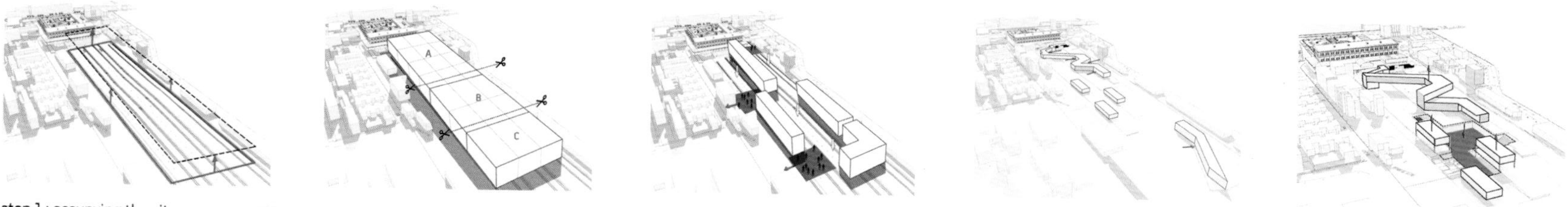

step 1 : occupying the site

step 2 : cutting the block

step 3 : creating some squares

step 4 : generating the shape

step 5 : modifying the design

场地平面展示图与设计演变过程图

URBAN DESIGN
RENOVATION PROJECT OF PUKOU RAILWAY STATION

历史街区城市再生设计
Historical Urban Area Revitalization

SECTION

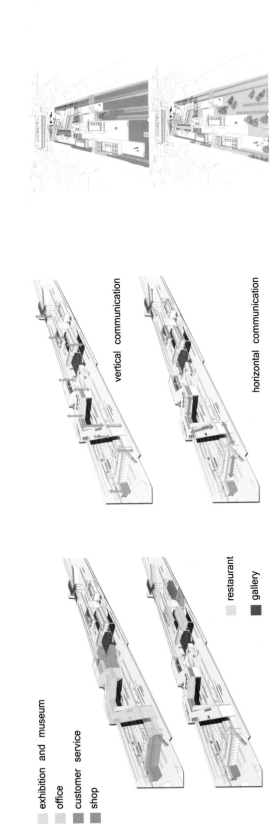

PLANTING AND LANDSCAPE ANALYSIS

vertical communication

horizontal communication

COMMUNICATION ANALYSIS

restaurant
gallery
exhibition and museum
office
customer service
shop

FUNCTION ANALYSIS

URBAN DESIGN
RENOVATION PROJECT OF PUKOU RAILWAY STATION

MOVEMENT THROUGH MAJOR ENTRANCES AND EXITS

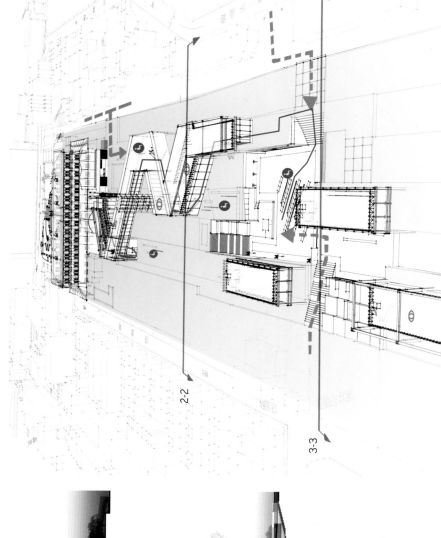

- rest zone
- viewing point
- main entrances
- pedestrian flow

SECTION 2-2

SECTION 3-3

In the whole design we try to use the height difference of platform to define the boundary and the entrance of the squares.

As the rendezvous point, the sunken plaza not only provides the vistors a space to rest, but also an atmosphere to distinguish the exhibition area from the leisure area.

局部剖面图与流线分析图

URBAN DESIGN
RENOVATION PROJECT OF PUKOU RAILWAY STATION

SECTION

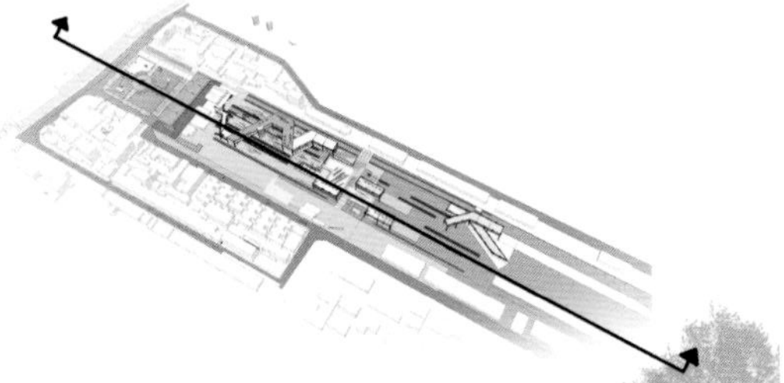

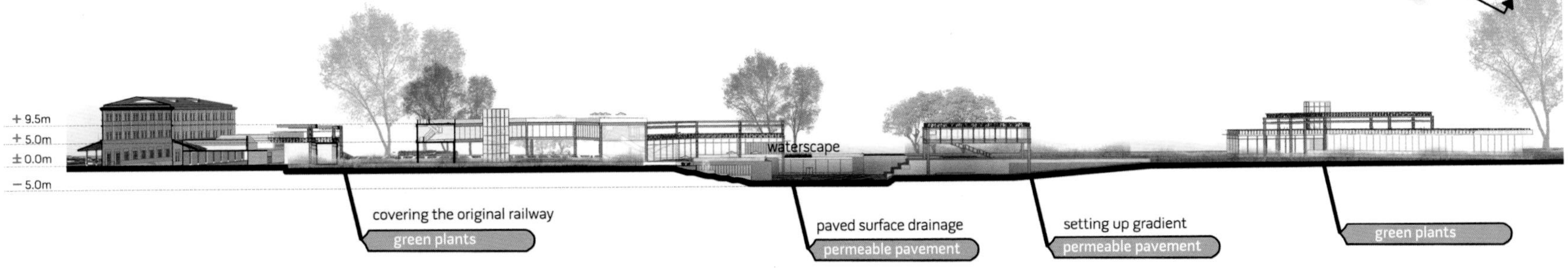

+ 9.5m
+ 5.0m
± 0.0m
− 5.0m

covering the original railway — green plants
paved surface drainage — permeable pavement
waterscape
setting up gradient — permeable pavement
green plants

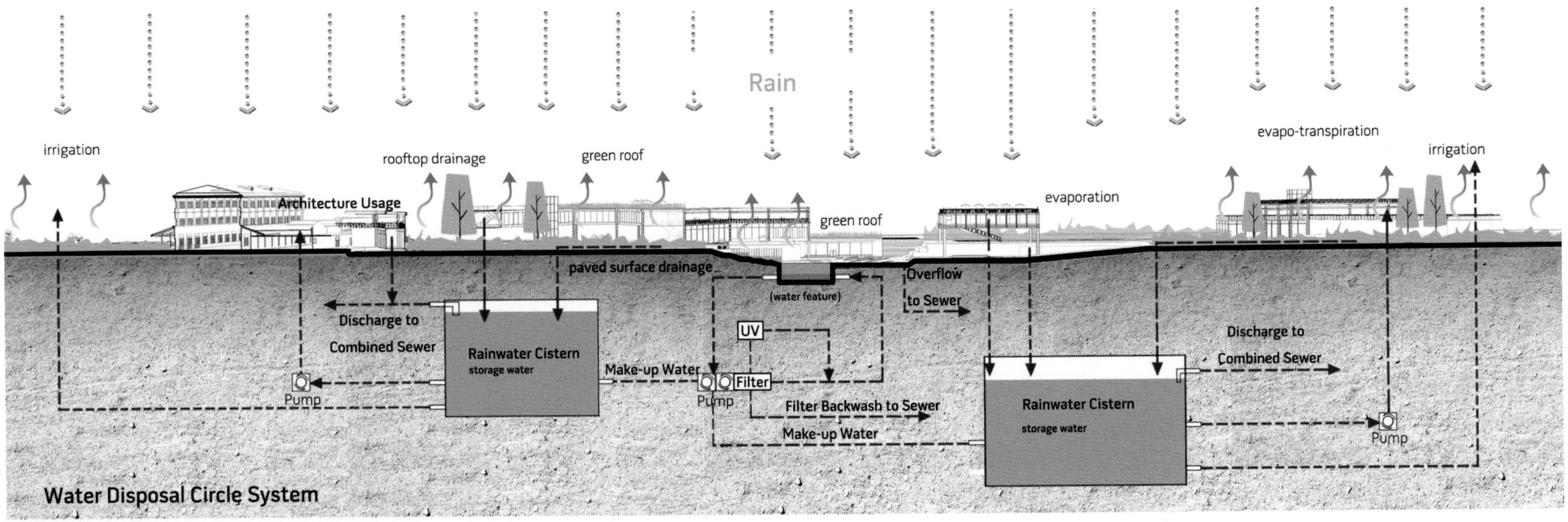

Rain

rooftop drainage
Architecture Usage
green roof
paved surface drainage
green roof
(water feature)
evaporation
evapo-transpiration
irrigation

Rainwater Cistern — storage water
Discharge to Combined Sewer
Pump
Make-up Water
UV | Filter | Pump
Filter Backwash to Sewer
Make-up Water
Overflow to Sewer
Rainwater Cistern — storage water
Discharge to Combined Sewer
Pump

irrigation

Water Disposal Circle System

雨水回收利用分析图

218

校园景观设计
Campus Landscape

三、景观类专业方案

科罗拉多大学校园外道路景观设计（米兰理工大学）

米兰理工大学建筑设计硕士 李炫静

Fort Collins, Colorado is a small city located at the base of the Rocky Mountain foothills along the Cache la Poudre River in northern Colorado with a rapidly-growing population numbering just over 161,000 at the start of 2016. Sitting at an elevation of 5,000 feet above sea level, the city enjoys a sunny four-season climate with mild winters. Currently in the midst of a cultural renaissance period, Fort Collins continues to frequent many "best-of" lists as a great place to live, work, and play, and was named Money Magazine's "Best Place to Live" in America in 2006.

03 OFF THE RAILS

Period : June.2016
Acedemic Year : Fourth
Location : Fort Collins, Colorado
Type : Competition | Individual Work

背景介绍与方案鸟瞰图

The largest employer in the city is Colorado State University (CSU), a major land-grant research institution and source of talent for the city's many high-tech employers such as Hewlett Packard, Intel, and Avago. Known as a craft beer-brewing mecca, Fort Collins boasts 20 award-winning craft breweries, New Belgium and Odell as trailblazers, along with a regional brewery for one of the best-known large beermakers in the world, Anheuser-Busch. Art and culture thrive in this city, which features a vibrant, historic Old Town district for shopping and dining, a half-dozen live theater stages in production throughout the year, and a flourishing local music scene. Recreational, competitive, and commuter cycling are also extremely popular in Fort Collins, rounding out its reputation as a great place for 'Beer, Bikes, and Bands.'

Architecture strategy

Public space strategy

Landscape strategy

Transportation strategy

Existing context

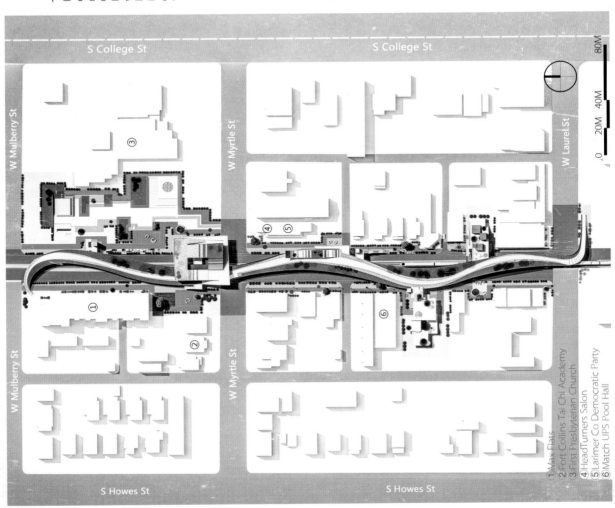

1 Max Flats
2 Fort Collins Tai Chi Academy
3 First Presbyterian Church
4 HeadTurners Salon
5 Larimer Co Democratic Party
6 Match UPS Pool Hall

校园景观设计
Campus Landscape

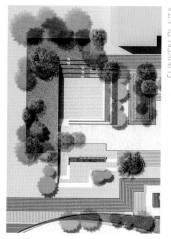
SUNKEN PLAZA

NODE I

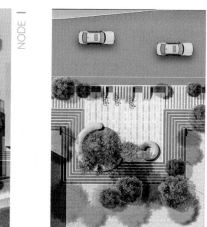
NODE II

The site is a 0.23-mile segment of the Mason Street Downtown Corridor from Laurel Street to Mulberry Street. Colorado State University, home to 32,236 students, bounds Laurel Street to the south. North to Mulberry Street supports a mix of residences and commercial establishments. The site serves as a connector, transitioning the University to the Downtown.

节点分析图及平面展示图

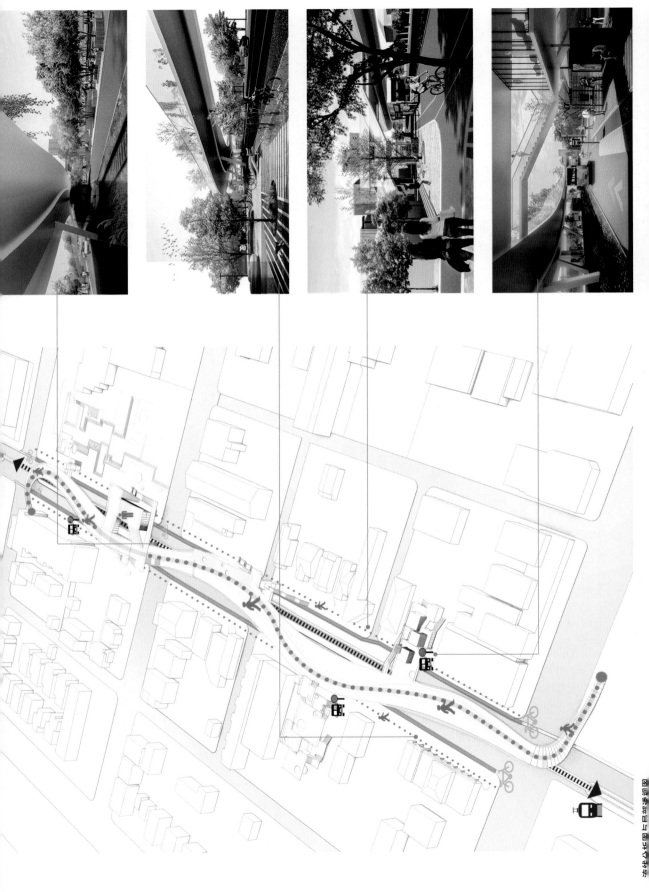

校园景观设计
Campus Landscape

Along this segment, multiple transportation modes converge. The site is well-served by public transport with the MAX line. At the Laurel Street intersection MAX joins the two-way street. Two stations are located along the site – Laurel station and Mulberry station.

节点透视图展示

The Mason Trail, a popular bike route for students and professionals, merges onto Mason Street at Laurel Street and connects to the city's larger route system. The BNSF Railroad track centrally divides the street as car traffic travels alongside. Pedestrians frequent the area to access its amenities.

主要建筑展示

Over the past decade the site has seen growth and change supporting high public use. Various redevelopment and infill projects have been spurred by the MasonCorridor economic development initiative anchored by the MAX BRT system. The site is transforming into a popular and progressive urban space for all ages and uses.

校园景观设计
Campus Landscape

SUSTAINABLE CAMPUS PEDESTRAIN ZONE DESIGN

ROAD DIMENSION COMPARSION

IN CONSTRAST WITH THE URBAN ROAD SYSTEM IN WESTERN COUNTRIES, ONE OF THE MOST PRESSING PROBLEMS IN CHINA IS THAT THE OVERSIZE STREETS TEND TO LACK OF INFRASTRUCTURE AND LANDSCAPE. THERE IS ALSO AN ABSENCE OF NON-MOTOR VEHICLE LANES.

ROAD NETWORK IN MUNICH

ROAD NETWORK IN QINHUANGDAO

ROAD NETWORK IN LYNOS

ROAD NETWORK IN NEWYORK

IT IS WIDELY ACKNOWLEDGED THAT FROM THE PERSPECTIVE OF URBAN DESIGN, THE IMPROVENT OF TRAFFIC CAN HELP ENHANCE CITY VITALITY. THIS IS ALSO VALID FOR CAMPUS SCENE, STARTING WITH THE MODIFICATION OF STREET SCALE, THE MATURITY OF ROAD CLASSIFICATION, AS WELL AS THE INCREASE OF LANSCAPING TO BEAUTIFY THE STREETS.THIS DESIGN AIMS TO ILLUSTRATE THE PROCEDURE OF TRANSFORMING THE BUSIEST CAMPUS STREET TO AN UNIMPEDED ONE WITH FIVE NODES.

CATEGROY: STUDIO COURSE ASSIGNMENT 4TH YEAR INDIVIDUAL WORK
TIME PERIOD: 8 WEEKS
LOCATION: QINHUANGDAO, HEBEI PROVINCE

PLAN ROAD NETWORK

=

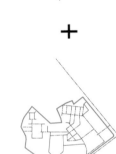
PUT IN ORDER

+

校园景观步行系统设计（米兰理工大学）
获得米兰理工大学可持续建筑与景观硕士 offer 朱立群

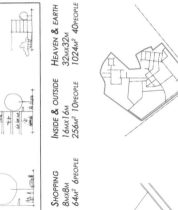

CURRICULUM THEME IS URBAN DESIGN
TRANSFORMATION BASED ON SCHOOL
TARGET SCHOOL:HEBEI NORMAL UNIVERSITY OF SCIENCE & TECHNOLOGY
LOCATED IN THE COASTAL AREA OF QINHUANGDAO CITY
AERIAL VIEW

ORIGINAL ROAD NETWORK

ORIGINAL LANDSCAPE

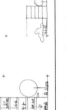

HEAVEN & EARTH
32M×32M
1024M² 40PEOPLE

INSIDE & OUTSIDE
16M×16M
256M² 10PEOPLE

SHOPPING
8M×8M
64M² 6PEOPLE

RELAXING
4M×4M
16M² 4PEOPLE

MEETING
2M×2M
4M² 2PEOPLE

WALKING
1M×1M
1M² 1PERSON

ROAD DIMENSION ANALYSIS

Teaching Area
Living Area
Dormitory Area
Sport Activites Area

ORIGINAL FUNCTION

All-round perspective

These five nodes are evenly distributed through the whole street. They can be found at the staring point as well as the central corner and so forth.

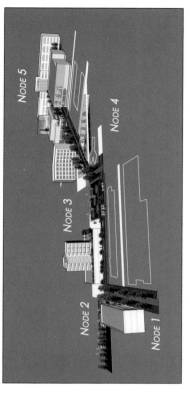

Node 1

Considering the fact that the first node is the starting point of the whole street, a square is added here. Such kind of spacious public area is expected to bring more active atmosphere to this place.

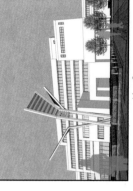

Note 2 perspective drawing

Master plan

Based on the questionnaire study, the street which traverses dormitory and living areas, together with two teaching areas is labeled as the one with the largest pedestrain volume. Thus, it becomes the main object of this research. Attention is fixed on the creation of proper street scale, as well as the proprotion of the building and energy-saving landscape node sets.

设计总平面图

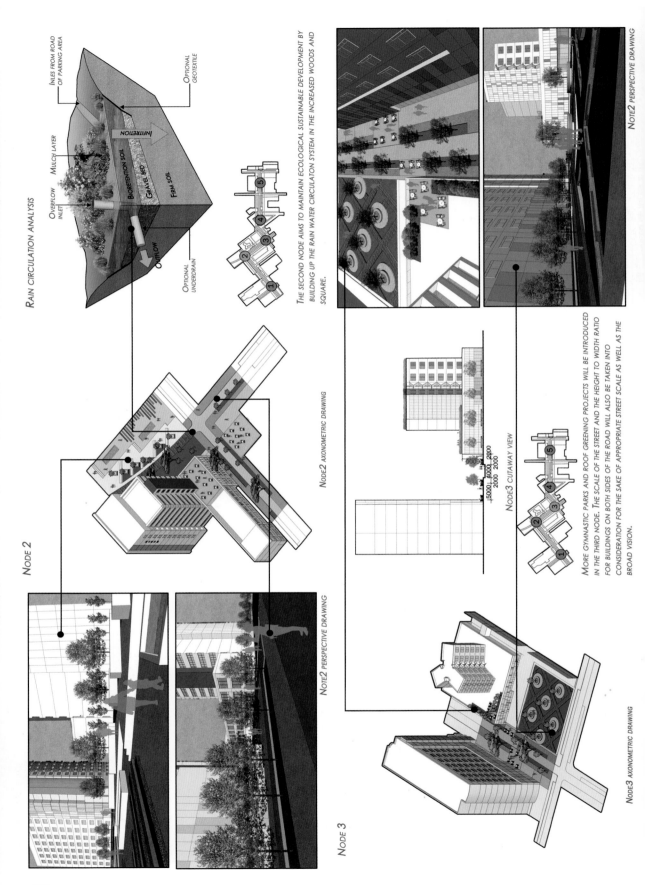

NODE4 PERSPECTIVE DRAWING

NODE5 PERSPECTIVE DRAWING

THE PEOPLE CAN FEEL THE GREENERY ILLUMINATED BY SUNSIGHT FROM EVERYWHERE.

THE FORTH NODE ATTEMPTS TO EXPLAIN THE FUNCTION OF SUNKEN SPACE AND ROOF GREENING PROJECT. THE MAIN FOCUS LIES ON THE VISUAL EFFECT OF SUCH PROJECT IN THE ROAD.

NODE 5

NODE4 AXONOMETRIC DRAWING

NODE5 AXONOMETRIC DRAWING

THE FIFTH NODE FOCUSES ON THE RAINWATER COLLECTION SYSTEM RUNNING THROUGH THE STREET.

RAINWATER COLLECTION SYSTEM

Once in the bioswale, some water infiltrates the ground, replenishing groundwater supply.

Unabsorbed stormwater is discharged into a subsurface drain.

During heavey rains, Excess water is channeled under the sidewalk grates into the bioswale.

From the filter strip, water is collected by an underground drain and is channeled into the bioswale.

BIOSWALE

FILTER STRIP

Rainwater runoff from the road into planted filter trip, then into the bioswale.

River stone slows down water flow before it enters the filter strip.

Suspended particles begin to settle out of the runoff with the help of the plants and soil.

节点分析及生态技术展示图

公园景观设计 / Park Design

公园景观修复设计（米兰理工大学）
米兰理工大学可持续建筑与景观硕士　王雨歆

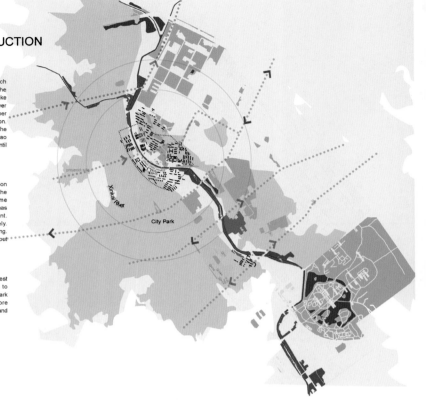

TANAN PARK RECONSTRUCTION

Background information

The park located in Shenyang, Liaoning province, which faces the Xinkai River in the south. The river cross the whole city and flows southeastward to the Dingxiang Lake because of certain altitude. The southern part of the river is densely-populated residential area, in contrast, the other side of which is undeveloped suburb living fewer population. The park caters for surrounding three communities. The Sheli Tower and a temple which were built during the Liao Dynasty in the park are famous for the local populace. Until now many residents often go to the tower for prayer.

EXISTING CONDITIONS

The water quality quite exist a lot qustion. Water pollution imperils all kinds of aquatic plants and animals. Thus, the ecological balance in the river may be disrupted. Some places is even choked up with silt. Therefore the park has become a necessary venue for public entertainment. Existing vegetation, like grass and shrubs are sparsely. Some old buildings are brocken and lack value of visiting. The small lake in the park is effectively independent without any connection with the Xinkai River.

PROPOSAL

Improve the quality of the water system. Increase forest coverage rate. Let people obtain more oppotunities to interact with water and close to the nature. Covert the park into a more characteristic. Then attract more and more people to invest reconstruct the old building in the park and develop the northern areas.

方案背景介绍及区域总体分析

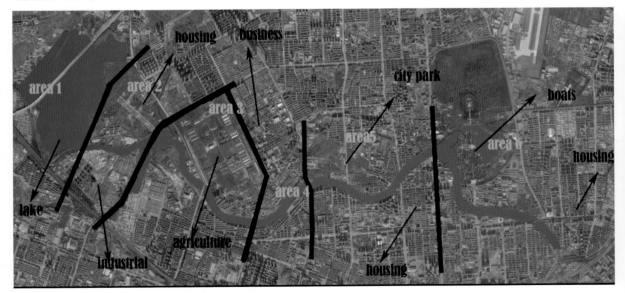

DIVIDED AREA
[ZONING]

Preserved area are divided into six areas by different functions. Area one, two and three are most fields, mills and lake. The lake is used to irrigate the whole agricultural areas. The rest areas have a higher proportion of housing and also commercial area. So the throng are rather denser. Project site is on area four.

设计区域功能分析图

公园景观设计
Park Design

TRAFFIC CONDITION

current condition

Five routes connect two sides of Xinkai River.The traffic in Road three near the project site is quite heavy and slow.Use of the road is also disorderd.

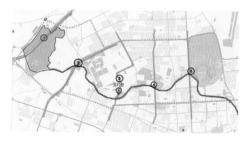

Design concept

To ease the traffic pressure,the square of the park can be used by pedestrain and runners.It is also an indirect way to widen the road,meanwhile,wide square can attract more pedestrains to visit the park or have a rest.

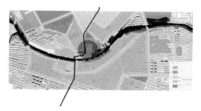

Bus&Parking

Bike&Pedestrian Collision With Automobiles

Heat Map(Vehical)

Heat Map(Pedestrain)

现状解析及设计策略

RIVER TRAIL

Current condition

The bad water quality of the existing river way can cause a lot of question,the major sources of water pollution are the waste from factories and cities.polluted water flows to the rural area have the effect on irrigation and living.People near the water are hard to contact the water because of pollution and water level.

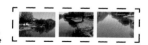

Design concepts

To Use some serried ranks of railings to intercept the waste and stones in river upstream Meanwhile, purify the water for irrigating the filed.Use several existing river cycle waer .

To Set two water gate at the housing area to rise water levels so that more people can interact the water.Then attract more people to rural area.

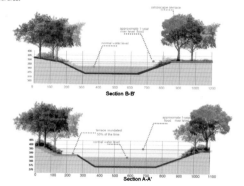

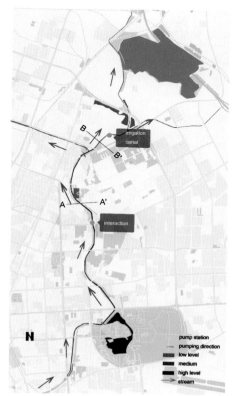

景观竖向设计分析图

公园景观设计
Park Design

REGIONAL RESEARCH

Surrouding research

The park mainly serve three residential quarters,and the fuctions of the park should cater to different kind of people ,most of them are faniles and the eldlerly.There are quite some recreation areas near the park but they can not meet the demand of closing the nature.

Design concepts

The location and characteristic of different fuctions should base on requirement of visitors

To Make full use of the only green area in the park and the small lake to create a natural and private space,

The fuctions should let all visitors feel comfortable no matter when

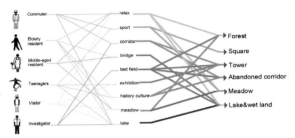

strategies: To Divide visitors into different types,and research their requirements. Finally,summarize six different functions

理念再阐述提升图

LOCAL HISTORY

Historical research

The temple and the tower are symbol of the park and nearby residents also maintain the tradition that bless themselves in tower.But the corridor connected the temple and tower is broken and faded.

The custom has been passed down since the 16th century.Taking full advantage of this tradition helps preserving the historical value of the park and developing the relevant industries which offers more job opportunities.

Design consept

- To Remain the outline of the broken corridor and the colour of pillars.The new one contrasted sharkly with the old one but let people konw what did it look like before.
- To Amplify the square of the tower,serve the historical function to more people
- To Make a short cut from gate to the temple and temple to tower for the people with definite destination.

People built this park only for historical value in the early year

Paroche house Moerenburg with it's formal garden was part of a sewery area until the 1920s.

New design: house was 'rebuilt'and the gardenstructure was remade into a helofytengarden and podium like the original function. This area was appointed as a water cleaning area

设计语汇提炼图

公园景观设计
Park Design

FUNCTION AND RECONSTRUCTION
Different location and their location
designed from visitors' habits and customs

Change
- Noise and obstacles are created by the street hawkers. The gate of the park should be clearly recognizable
- Barren lands are exposed at the entrance Utilize bushes separate roads and the park
- Bad connections to surrounding transport infrasructure.
- Danger slopes near the lake with big sharp stones
- No terrace or flat for visitors.Less entertaining.

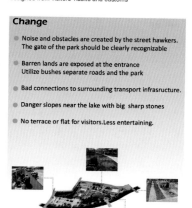

"Design is for peoople,serving people's life demands are the final purpose of the design."

- To Retain existing forest,but more elbowroom should be provided.
- To Extend the square of the tower.
- To Set up the entrance on the right of the park which is easy for the residents to go to.

- Rich and attractive landscape plants.
- Aquatic plants mainly with terrain target. Provide the observation near the water.
- Large area of lawn for visitors to have a rest.

- To Connect the lake and the river so that water in the lake flows to the river because of altitude
- The lake is expected to collect rain from the whole park,this is a water circulation process

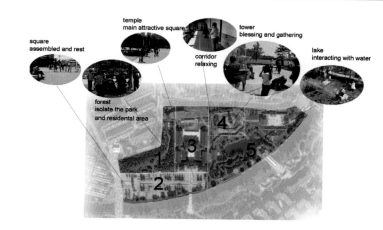

- square assembled and rest
- temple main attractive square
- tower blessing and gathering
- corridor relaxing
- lake interacting with water
- forest isolate the park and residental area

功能再植入设计图

Master plan

Plants
- higher bushy trees are planted at the edge of the park near the park to isolate noise and dust.
- A large area of glass where people can lie down in glass to have a rest.

Axis
- Each axis runs through a historic building of the park.
- Each axis connects the bridge which is main road for people th visit from the other side.When people walking on the bridge,they can always see the symbolof the park.

Layout
- People can close to the nature at the right of the park.
- Bushes are planted near the water,the density of them decides diffferent sight lines and visual field.People stand on different place can see different views.

when people walk from square to forest. the row of higher trees seems like a"door"as a transition.

When people stand at the entrance,they just can the whole scenery.if they want to see more,they have to round the lake so that enjoy different landscape

section A-A"

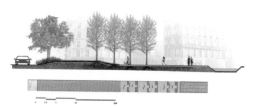

section B-B"

设计总平面图与景观轴线分析图

公园景观设计
Park Design

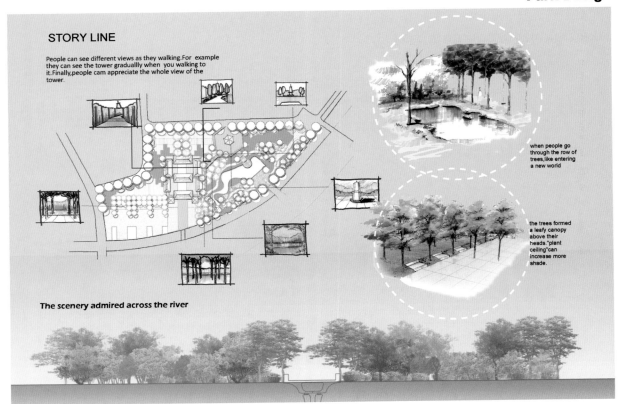

景观节点展示图

景区效果图展示图

城市宗地景观改造
Urban Brown Field Landscape Design

沈阳工业区景观再设计（米兰理工大学）
米兰理工大学可持续建筑与景观硕士　王雨歆

INDUSTRAL ABANDONED LAND LANDSCAPE

BACKGROUND INFORMATION

Shenyang is one typical industrial city in China, because of the role she played in China's social and economic development history. At the same time, Tiexi district is the industrial heart of Shenyang. The goal for the project was to renovate the landscape and future design of an urban industrial abandoned land, which was located in Tiexi, Shenyang. The whole abandoned landscape is 27 hectares and the project occupies 20 hectares. The Cast Museum of China and Weigong open channel are right to at the west of the project, and the east of the project are huge areas for residence and education institutions. More than that, it is also at a cross-point in Tiexi's lounge road which bring it great advantages in traffic convenience, rich culture, abundant resources, and typical local characters.

DESIGN IDEA

To keep the original inner buildings of the abandoned landscape is the core of the project. Imagining that in our daily lives, we have been hearing too much noise from the concrete jungle, like yelling from selling and loudly car horns from angry drivers. While how long we have not paid attentions to the sound of footsteps, lovely children's chuckling and gentle wind.

设计背景及理念阐述

SOUND SOURCE & TRAFFIC

Imagining that in our daily lives, we have been hearing too much noise from the concrete jungle, like yelling from selling and loudly car horns from angry drivers. While how long we have not paid attentions to the sound of footsteps, lovely children's chuckling and gentle wind.

We devided sounds into human sounds, natuural sounds and human&natural sounds.Those sounds come from students in school,machine in factory, trees, brids and so on. So when we walk on the trail in the site we can feel different sounds.

Design Issues

▲ Industrial abandoned land completely to save and reuse.

▲ Existing landscape of the site innovation and reuse

▲ Open space in series with the activation

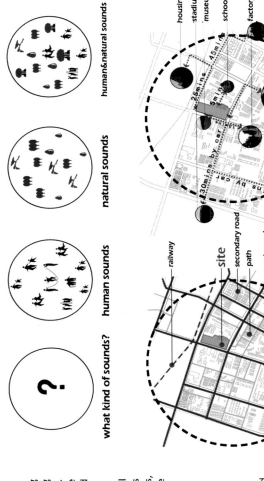

Site Sound Analysis

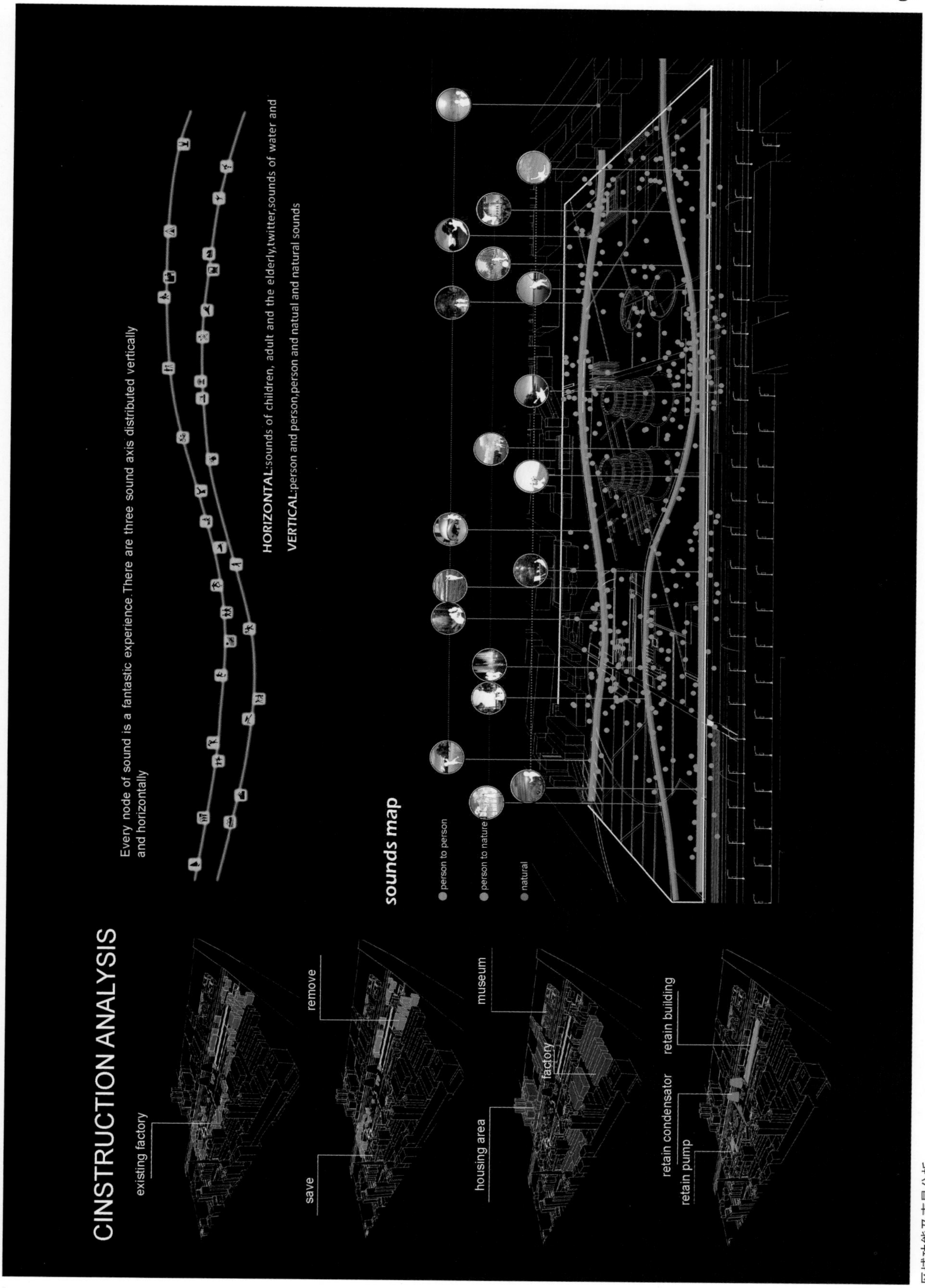

城市宗地景观改造
Urban Brown Field Landscape Design

CINSTRUCTION ANALYSIS

existing factory

remove

save

museum

housing area

factory

retain building

retain condensator

retain pump

Every node of sound is a fantastic experience. There are three sound axis distributed vertically and horizontally

HORIZONTAL: sounds of children, adult and the elderly, twitter, sounds of water and
VERTICAL: person and person, person and natural and natural sounds

sounds map

- person to person
- person to nature
- natural

区域功能及声景分析

MASTER PLAN & CONCEPT

Without the extensive attention, the modern industrial monuments are vanishing gradually under currently large-scale new buildings inrushing. Even though to protect the industrial monuments is always the core concept for some designers recently, eventually few of them got great success. While the project in **Ruhr, Germany** is an excellent example in the industry. So for the designers, we should pay more attentions on the industrial monuments protections to last the proud history and culture of our home country. Particularly for Shenyang who has been famous for an industrial city for a long time, there are more meanings and responsibilities for us to learn more about the design ideas and methods to survey the industrial monuments.

The design reserved most of the inner buildings and structures of the abandoned industrial area to convey the value of **"at present"**. On the basis of that, a continuous landscape is being completed. Under the study of different functions of sound, **"Sound"** is the cut point and charming part of the project making the whole project user friendly and attractive. The new utilities design, reasonable space planning bring tremendous successes to the typical landscape creation and industrial culture inheritance. With the reasonable buildings and landscape's rearranging and needs from the modern life, a new typical culture strict is given to birth with abundant culture background.

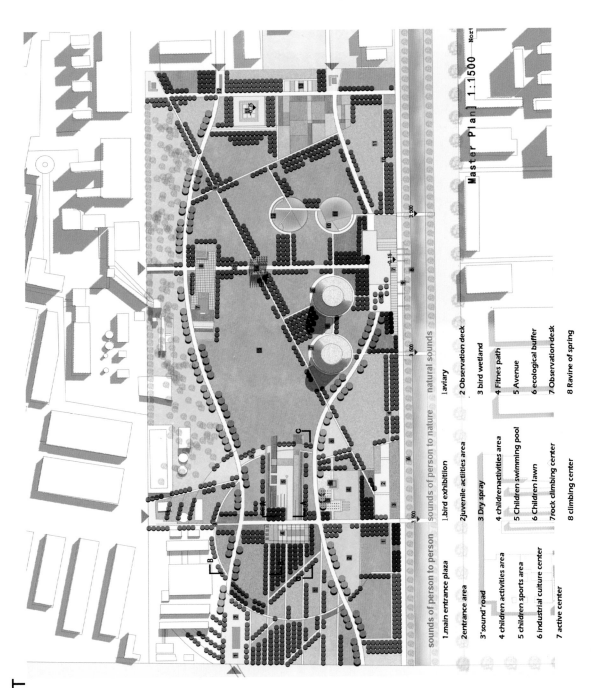

sounds of person to person	sounds of person to nature	natural sounds
1.main entrance plaza	1.bird exhibition	1 aviary
2 entrance area	2 juvenile actiities area	2 Observation deck
3 'sound' road	3 Dry spray	3 bird wetland
4 children activities area	4 children activities area	4 Fitnes path
5 children sports area	5 Children swimming pool	5 Avenue
6 industrial culture center	6 Children lawn	6 ecological buffer
7 active center	7 rock climbing center	7 Observation desk
	8 climbing center	8 Ravine of spring

Master Plan 1:1500

设计总平面图与设计理念

城市宗地景观改造
Urban Brown Field Landscape Design

ENVIRONMENTAL ANALYSIS

Roads | Cinstruction | Vegetation | Water

Weaver — Swida alba aopiz
Cypress
Codonopsis
Loriot — Weigela Primroses
Peach
Throstle — Forsythia
Magpie
Siskin — Ginkgo
Albizzia

景观体系构成图及元素分析图

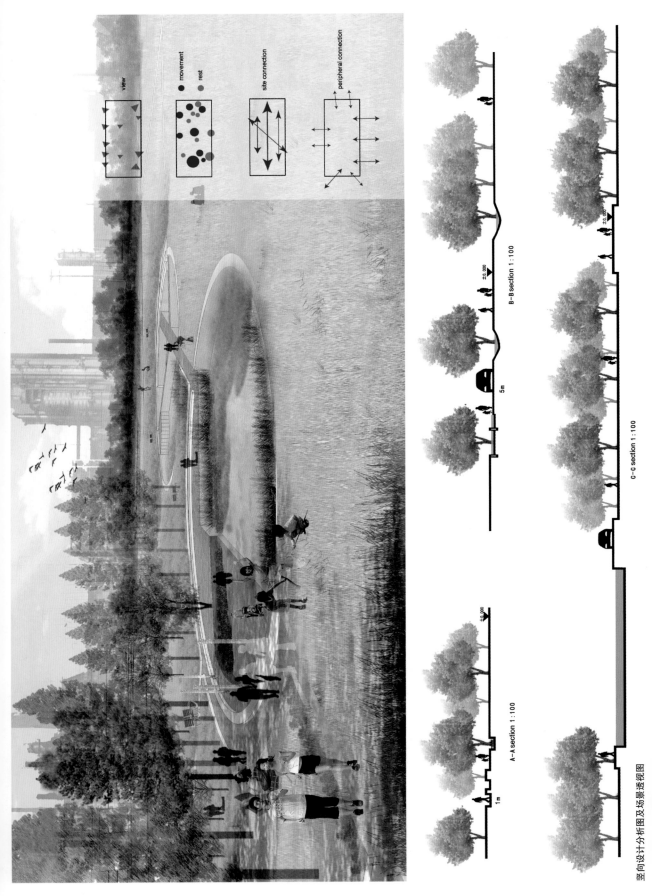

竖向设计分析图及场景透视图

休闲景观设计
Leisure Landscape Design

辽宁本溪温泉景区景观设计（米兰理工大学）
米兰理工大学可持续建筑及景观硕士 王雨歆

LIAONING TANGGOU HOT SPRING RESORT

BACKGROUND INFORMATION

The site is locates in Caohezhang Town,Benxi County,Liaoning province,the bade has a very superior geographical conditions and location advantages.Benxi county has a total area of 3344 square kilometers and features as the "Eight mountain land a monohydrate",it belongs to the north temperate zone continental monsoon climate. The county jurisdiction over 11 township and a total population of 300000 people.The terrain of Benxi is the southeast part of Changbai mountain. The terrsin is high in southeast,and low in northwest.There are morthan 240 mountains.

Main mountain ranges are above 1032 meters

EXISTING CONDITIONS

Nowadays people are living in the fast paced environment.Peaceful area in countryside is required.Samanism is the traditional Chinese culture originated in Benxi and the special cultural attracts a lot of peolpe .Tang Gou infernational hot spring resort tourism resources can be divided into natural and humanistic categories,giving priority to natural resources,accounting for 67％.Natural resources are divided into four categories,physiographic landscape,water scenery,biological landscape astronomical phenomena and climate. human resources are two main classes,travel products and cultural activities.

PROPOSAL

The rough montain tracks could be reconstructed into hiking tracks for people to close to natua and escape from modem life.we would like more and more people love Saman culture. People would change their lifestyle when they live there.

背景及设计愿景阐述

休闲景观设计
Leisure Landscape Design

RESEARCH

The site is located in Chaohezhang Town, Benxi county, Liaoning province, which is the cradle of some of Chinese richest civilizations. The site is surrounded by hills, its nature and culture attracts a lot of people.

Benxi manchu autonomous county is the national famous tourist resort, also known as "YanDong scenic spot". There are six national scenic areas.

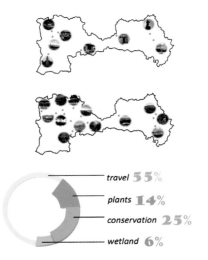

travel **55**%
plants **14**%
conservation **25**%
wetland **6**%

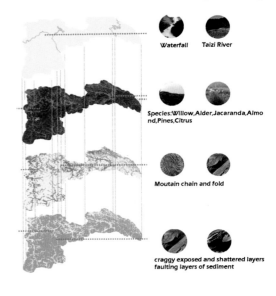

Waterfall Taizi River

Species: Willow, Alder, Jacaranda, Almond, Pines, Citrus

Moutain chain and fold

craggy exposed and shattered layers faulting layers of sediment

ROCK: SANDSTONGE, MARL, GRAVEL, CONGLOMERATS

调研分析图展示

HIKING TRAIL

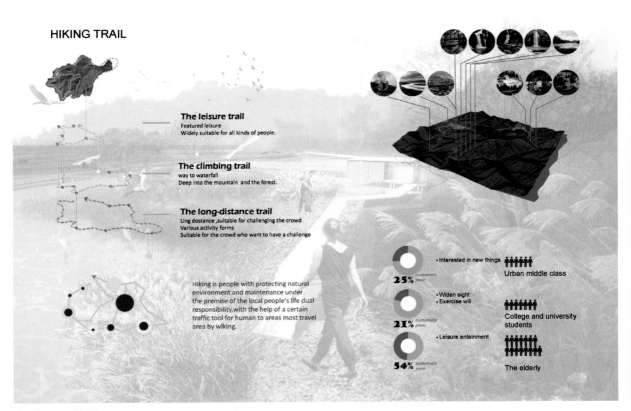

The leisure trail
Featured leisure
Widely suitable for all kinds of people.

The climbing trail
way to waterfall
Deep into the mountain and the forest.

The long-distance trail
Ling dostance, suitable for challenging the crowd
Various activity forms
Suitable for the crowd who want to have a challenge

Hiking is people with protecting natural environment and maintenance under the premise of the local people's life dual responsibility, with the help of a certain traffic tool for human to areas most travel area by wlking.

- Interested in new things
25% customers pose
Urban middle class

- Widen sight
- Exercise will
21% customers pose
College and university students

- Leisure entainment
54% customers pose
The elderly

涉及人群调研分析图

休闲景观设计
Leisure Landscape Design

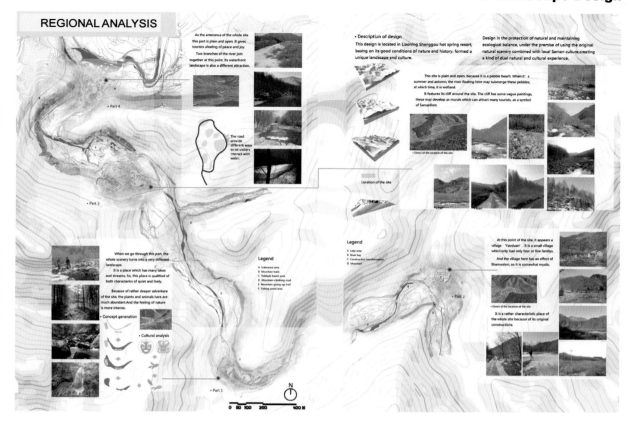

区域自然条件分析图

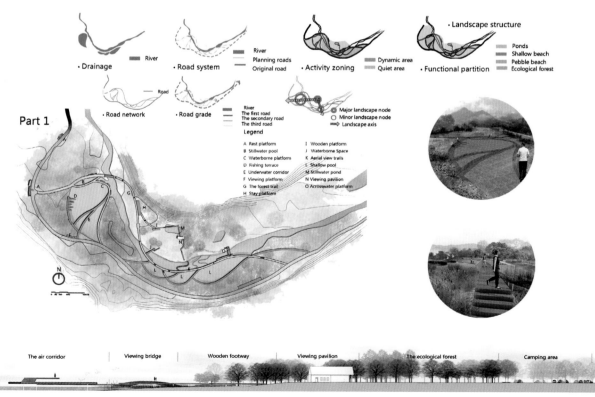

区域1景观体系设计图

休闲景观设计
Leisure Landscape Design

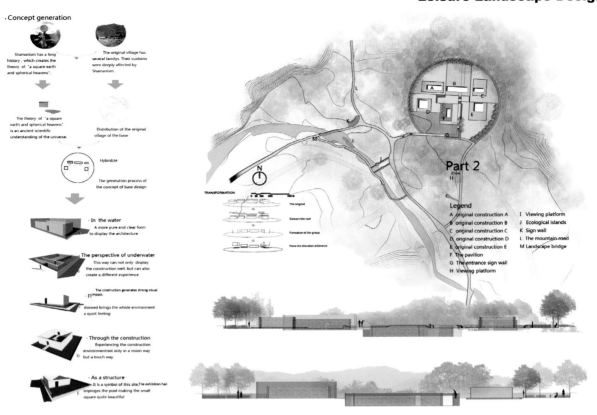

区域 2 理念展示及剖面图设计

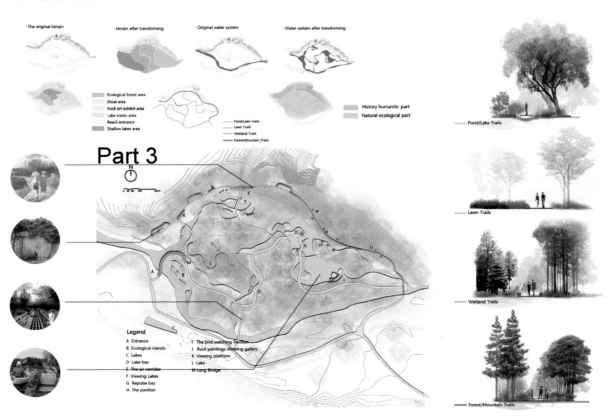

区域3 设计成果图展示

米兰理工可持续建筑及景观硕士　王雨歆

休闲景观设计
Leisure Landscape Design

THE PARTICULARITY

Thw particularity of the site is because of the contradictoriness.the modern facilities broke the texture of its natural village.The site is one of the village and now it is in the range of modernnization.A large number of agricultural areas have been reduced, which highlights the importance of protecting the agricultural civilization in the modernization drive.Before the founding of the PRC in 1949, this is still a piece of the concept of the village which is not developed in the community groups.After 30 years of development, there is an increase in the number of population, the site appeared the public gathering entertainment space, agricultural production and distribution in the village .After 30 years, a scale of the village has been formed, there is now a village with the farmland ecosystem and the spiritual beliefs of the village, the village continued to develop.Also experienced 10 years or so, the shape of the village to the direction of water development, the village population has been basically maintained, people, animals and nature to maintain a stable ecological relationship.

1947　1956　1977　1987

The urban fringe areas have a lot of factories, which affect the output of the farmland and the natural environment of the natural village.The city's living garbage is placed on the edge of the city, and the environment pollution disturbs the normal ecology.

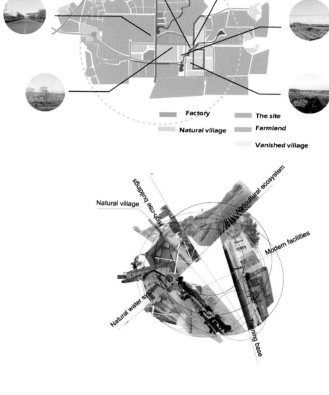

Factory	The site
Natural village	Farmland
	Vanished village

设计背景分析

DESIGN CONCEPT

From a macro perspective

From a macro perspective, the village of general form elements have mountain, water, land, village four necessary elements.Mountain, water, land, village four necessary elements, firmly inseparable.Since ancient times, the Chinese people have lived in a social way, so the buildings in the village have been concentrated in the center of the village. The farmland system is distributed around the village. Farmland and water always have to be interlinked, and the construction of the mountain in line with"Face to the water,Back to the mountain"in Geomancy.The site could be characterized by feelings of local conditions and customs, developing tourism, promoting economic development

From a micro perspective

The llink of private yard and public space
The relationship of the layout of buildings and agriculture land.
According to the needs of different people, create a multi-functional tourist resort

Survey shows that in the list of occupations, the work of a large group of responsibilities are more like agricultural sightseeing tours

The survey shows that agricultural tourism is popular among young people and middle-aged people.

In farmland for basal, architectural form dot scattered layout, each store aggregation form residential groups and courtyard

The open space provides people a place of entertainment

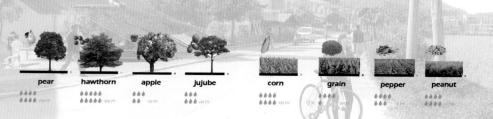

pear　hawthorn　apple　jujube　corn　grain　pepper　peanut

设计理念和植被分析

休闲景观设计
Leisure Landscape Design

MASTER PLAN

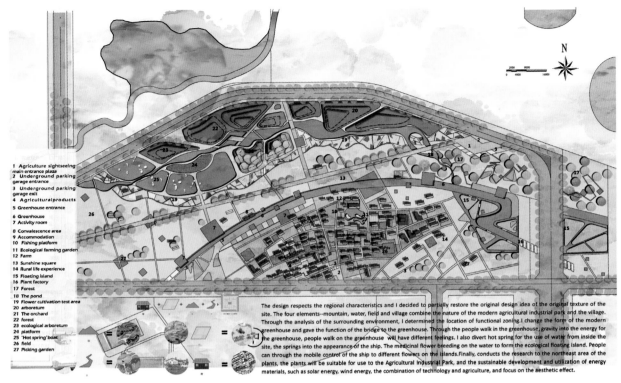

设计总平面图展示

FUNCTION
Mountains, water, field and village are endowed different fuctions

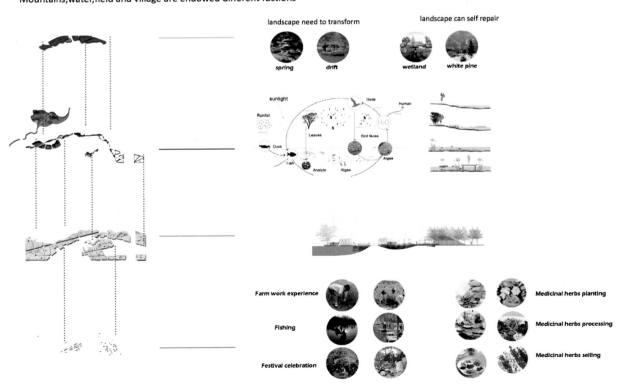

景观功能分析图

休闲景观设计
Leisure Landscape Design

ANALYSIS

The ship conbined with hot spring,which can let visitors enjoy the spa and appriciate different scenery. By moving the ship,people can pick flowers for spa.

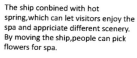

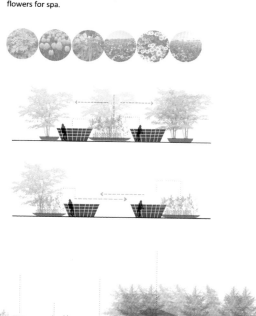

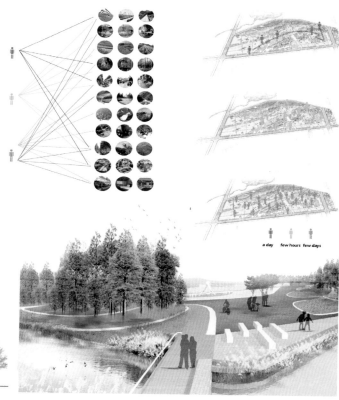

场地使用分析图及人群活动特征分析图

Greenhouse

Walking on the Greenhouse

Ecosystem circulation

Energy conversion diagram

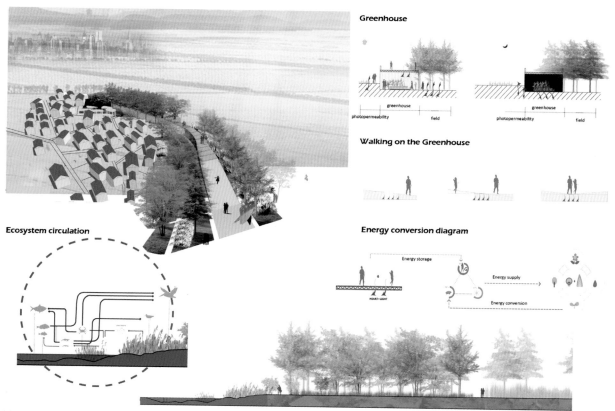

循环生态景观系统设计及节点大样图

245

四、室内设计类专业方案

住宅室内设计
RESIDENTIAL INTERIOR DESIGN

公寓室内设计（林肯大学）
获得林肯大学室内设计硕士 offer　王彬琪

APARTMENT IINTERIR DESIGN

Period: March.2014-June.2014
Type: Academic Work
Location: Suihua, Heilongjiang
Academic Year: 1

Themed with modern simplicity, this residential house is suitable for the family of three. Focusing on concise and vivid style, various elements featuring simplicity are integrated with each other. It not only emphasizes on the practicality of the house, but also it has embodied the personality and exquisite features of the modern life.

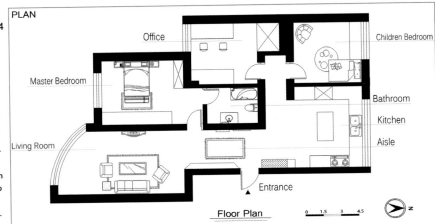

Space analysis graphics is drawn according to the "Flowing Space" by Ludwig. The white box in the middle represents toilet, which is regarded as service area. And the surrounding areas are considered as those receiving services.

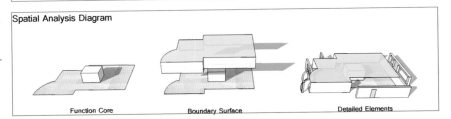

设计平面图及功能简介

Oak is the major decorating material the house adopts, matched with simple colors of white and black. The beauty and elegance of oak go flexibly with black and white, fully showing a sense of modern simplicity.

Bathroom has adopted water-proof floor, making cleaning and wiping more easily. The bathtub, toilet and basin are made of white ceramics. The bathroom is equipped with white spotlights to provide soft and bright illumination.

A cosmetic mirror has covered a large area of the wall. With specular reflection theory, the toilet seems to be enlarged visually. An independent sunlight spotlight on top of cosmetic mirror is designed for a supplement.

With greater humidity, some plants which are fond of moist are placed on the wash basin. The wet environment will moisturize the plants and boost them flourish, adding more vitality to the bathroom.

Dining room is the key part of the residential house, which not only requires the good-looking but also the integrality and practicability. On one side of kitchen are low cabinets. On the other side of kitchen are inlaid cabinets, which can save space and ensure convenience. Most kitchen wares are made of metal. Metal, as the products of industrial society, is important decorative element to reflect style of simplicity.

卧室、厨房及卫浴空间设计展示

住宅室内设计
Residential Interior Design

Studio Interior Rendering

Study room is a place for book storage and reading. With well designed daylight illumination, people visiting here can overlook out of the window when they are tired. Sofas are for people to take a rest and some plants are placed to relieve people from intensive reading.

书房设计展示

Living Room Interior Rendering

Living room can represent the taste of the owner. Oak wood floor, black sofa and cabinets, together with several pieces of simple pictures hanging on the background wall, creating a simple and graceful living room.

起居室设计展示

餐饮空间室内设计

哈尔滨中东铁路住宅改造——咖啡厅设计（威斯敏斯特大学）

获得威斯特斯特大学室内设计硕士 offer　王彬琪

COFFEE SHOP & BOOKSTORE INTERIOR

Period: March.2016-Junly.2016
Type: Academic Work
Location: Harbin, Heilongjiang, China
Academic Year: 3

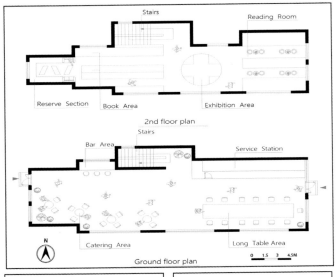

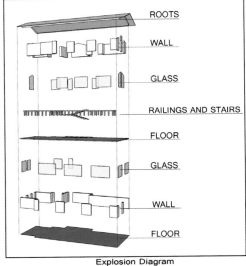

设计平面图及建筑结构爆照图

Material Funiture

Stools in the service area.

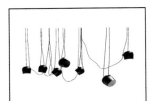

Ceiling lamp hanging above the long-table area

The stools in the long-table area

The original building is located in the city of Harbin, an area facing the street. It is a very valuable architecture that carrying lots of historical memory, therefore, it is very meaningful to preserved its historical facade and integrate modern functions inside to revitalize it. In this design, it is changed in to a pleasant cafe together with bookstore function upper floor.

Stepping upon 4 steps then the main entrance is presented, where you will first see the service area on your right side. In front of the service counter there is an exhibition area.
After some steps, you will come to the long-table area, the steps bring interesting visual experience to customers.

View Of Coffe Area

咖啡厅设计效果及设计元素展示

餐饮空间室内设计
Coffee Shop and Book Store Design

Material Funiture

Leather sofa

Round table for four

Book shelf

when the customers enter into the building, they are able to see the area of four people tables and the wine bar counter.

The floor has been altered to concrete floor; the lighting method of the chandeliers with three colors of lights is adopted to create the atmosphere; the large French windows enable the customers to feel the pace of the city and enjoy the charm of the sunset.

The are of four people tables is equipped with metal shelves, green plants and books are placed on the shelves to make the coffee house full of vitality.

The coffee house has two entrances/exits, the customer flow and staff flow are separated clearly.

View Of Coffe Area

咖啡厅设计效果及设计元素展示

Material Funiture

| The barstools in front of the bar counter | The ceiling lamps which can switch among three different colors | The table designed for four people | The simple geometric frame chair |

The cafe is equipped with a small bar counter. The floating space design enable customers to experience the dynamic activities in different corners of the cafe and feel the intimate atmosphere.

The whole cafe is industrial style, presenting mysterious and relaxing atmosphere by using black metal and wooden material.

 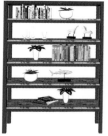

Details View Of Bar Area

吧台区设计效果及设计元素展示

249

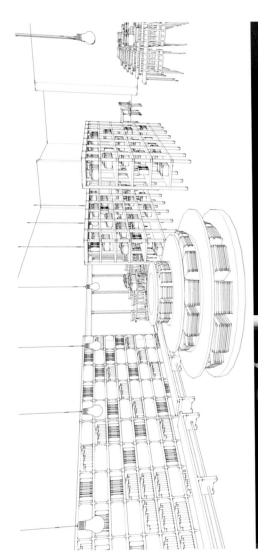

Study Area View

Go upstairs to the second floor where you can see the bookstore. It is divided into three areas: book displaying area, reading area and self study area. In the center of the bookstore, there is a round table to exhibit a wide variety of books and promote best-selling literature.

In the book displaying area, various kinds of books are placed according to categories, meeting the demands of certain groups. Some well-designed books or objects are put on the wooden shelves to attract customers. The ceiling lamps above shelves will ensure abundant warm yellow light and make people feel cozy with the sweet atompshere.

The bright and warm light in self-study area is one of the major light sources in the store. There's a small blackboard hanging on the wall, people who stay here can write the words to themselves or the their friends.

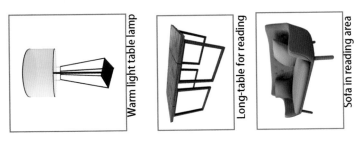

Warm light table lamp

Long-table for reading

Sofa in reading area

Octagonal table

Back-rest chair with four legs

幼儿园室内设计（伯明翰城市大学）
获得伯明翰城市大学室内设计硕士 offer　王彬琪

KINDERGARTEN INTERIOR DESIGN

Period: May.2015-August.2015
Type: Academic Work
Location: ShangHai, China
Academic Year: 2

Kindergarten is the important space to cultivate the young generation. systematic and scientific education in kindergarten is the important education beginning for young kids. Therefore, an architectural environment providing pleasant physical and mental environment for children is very essential.

The interior design indoors requires both diversity and simplicity. The spatial design of each classroom in the kindergarten is different. The classroom has a cheese-like shape, with different sizes of round shaped windows on the exterior walls. With sunshine going inside from different angles, various shadows will bring much fun to the children.

The gray interior wall is decorated with orange-yellow color, which will lead to a strong visual shock contrasted by the wooden floor.

Perspective View Of Classroom

Bedroom

Classroom Corner

幼儿园设计理念及儿童教室、休息室设计效果展示

Shared Activity Zone

Shared Activity Zone

Indoor Facility Elevtion

The public area in the kindergarten is designed as a playing district for children. With spacious room, this area is connected to the outdoors, where parents can easily send and pick up their children, as well as spend time playing with them.

Orange-yellow is the major color theme in the public area. Decorated with cartoon wallpaper, the entire area seems attractive. Small arc-shaped stones are paved on the floor for children's fun while protect their safety. The beautiful and practical chair can also serve as a square bookshelf.

A large children's slide is placed in the public area. Meanwhile, pretty chairs are provided in the surrounding area where parents can take a rest.

活动室室内效果及家居设计展示

第二节 不同国家院校作品集的要求要点（境外一流名校）

每个院校都有自己的作品集要求。同学们准备的时候需要按照学校的要求来。下面我们给大家列举一下主要的世界设计名校的作品集要求（2016—2017年度要求），给大家做一个参考。

一、英国

伦敦大学学院作品集要求

伦敦大学学院建筑系各个专业的作品集要求一般由学校通过邮件直接发给文书通过审核的申请者，没有通过初审的同学不会收到作品集提交的邮件邀请。以下是我们截取的邮件中关于作品集的要求陈述：

We have now received your form in the department and would like to request your design work. We require a printed copy of your work. This should be no bigger than A4. Please be aware that this is meant to be a representative summary of key design projects not a comprehensive record of your work.

Only send a COPY of your work as it will not be returned to you. Please do not send CD's, memory sticks or email attachments.

Please do not mount or print your design work on thick card; despite being very heavy and difficult to store it will be more expensive to post.

Clearly mark your design work on the front cover (not the envelope) with your full name, the programme you have applied for and your ID number and send it to:

Thea Heintz
Teaching and Learning Officer
The Bartlett Faculty Office
University College London (UCL)
22 Gordon Street
London WC1H 0QB

翻译过来伦敦大学学院对作品集的要求很简洁，同学们可以有很多自由发挥的余地。只要满足以下几点即可：

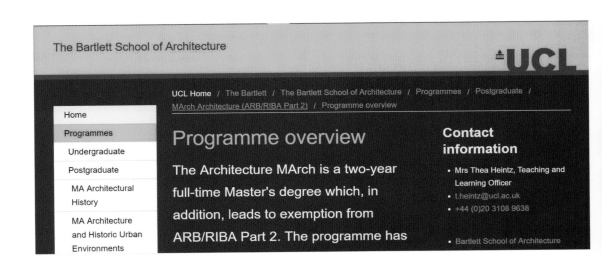

（1）作品集需要打印出来。

（2）尺寸不大于A4。

（3）包含最重要的设计而不是所有的设计。

（4）作品不退回，不要邮寄CD，U盘记忆棒，不要用邮件形式发送。

（5）不要用过厚的纸张打印，不要将作品加固在厚纸板上（过于沉重不便储藏）。

（6）在作品集封面（不是信封上）清晰标注你的全名，申请的项目名称以及申请ID号码。

因此同学们只要提前准备好A4的作品集即可，等到伦敦大学学院的秘书发来邀请邮件即可寄出。伦敦大学学院的老师都非常友善，从教学秘书到教授。曾经有教授看过同学的作品和材料之后主动建议调整申请专业。并且接收邮件形式发送新的作品集（特殊情况）。不过请大家寄出作品集之后耐心等待一周左右，如果2周还没有收到确认邮件，可以发邮件询问伦敦大学学院的秘书是否收到作品集。

谢菲尔德大学作品集要求

谢菲尔德大学建筑系对作品集要求如下（2017）

You will be required to submit a Design Portfolio. The following is some advice on putting together your portfolio:

- You should include a range of academic and practice work
- Use this as an opportunity to show your breadth and range of experience
- We do not need every drawing from every project
- Feel free to include extra activities i.e. competitions, workshops, fieldtrips etc
- This should be 10-12 sheets
- Please keep file size for this to a max of 20Mb
- You should submit this as a pdf. We do not accept links to online portfolios or shared drives.

谢菲尔德的作品集是在线提交，因此需要做成PDF格式。具体要求如下：

（1）作品集需要包括学术和实际工程作品。

（2）展示你知识的深度以及经验的广度。

（3）不用每个作品都事无巨细地展示。

（4）可以加入竞赛，workshop作品以及参观考察材料。

（5）总页数10到12张。

（6）PDF大小不超过20Mb。

（7）只接受PDF格式，不接受在线作品以及网盘分享等形式。

爱丁堡大学作品集要求

Portfolios are as diverse in style, content and format as the authors who assemble them. We value this diversity and so try not prescribe too heavily the format of the portfolio. The following are the basic parameters:

All work should be contained within a single PDF file, at appropriate resolution for screen view (hard copy/paper portfolios are not accepted)

Three-dimensional models, video, performance, or installation work should be represented in good quality photographs. We understand that many kinds of work are not easily represented within the portfolio format, so you will need to translate as best you can three-dimensional and temporal work onto the page.

Date all work and present it in a consistent way—thematically, by project and/or chronologically.

Include your name and UUN on the front of the portfolio.

Keep in mind that the portfolio will be used to assess the potential you have to benefit from and thrive within the programme. It is important, then, to look through this website to get a feeling for the range and character of design work being carried out at different levels the Department of Architecture.

Focus

The focus of the portfolio should be on your best and most recent work. It should include representative examples of work undertaken as part of a formal programme of study, and any work carried out while in practice, and/or self-initiated projects undertaken outside formal studies. As much as possible include a variety of work:

Kind—architectural design, speculative projects, art, built, research

Scale—furniture, buildings, urban design, regional studies

Media—freehand drawings, technical drawings, computer drawings, models, sculptures, live performances, paintings, installations, video

It is important that your portfolio contains your best work. But do not limit this to presentation images alone. Include sketches, studies and working drawings. If possible, include examples of from your design sketchbooks and notebooks.

Not all undergraduate courses in architecture, art and design require students to keep sketchbooks or notebooks, so you may not have such material to hand. However, notes and sketches play an important part in design thinking and are important for demonstrating your capacity to investigate and explore in the medium of design.

It is also important to present your most recent work. But, if you have had a longer period in practice or have a diverse educational, professional background, you should include work that would best show the wider scope and development of your career.

作品集要求简述：

（1）版式自由选择。

（2）提交电子版的 PDF 文件。

（3）合理地安排作品顺序，按照类型或者时间顺序。

（4）作品集封面上标注姓名和申请号。

（5）提交最近和最好的作品，尽可能展示你的能力。

（6）尽量多的作品类型，涵盖不同的设计规模。

（7）多样化的表达方式，手绘、计算机制图、模型、雕塑等。

曼彻斯特大学作品集要求

曼彻斯特大学和曼彻斯特建筑学院联合办学，申请建筑学方向的学生可以获得双学位。

作品集英文要求：

A digital portfolio of work (minimum of 35 pages) should be submitted with your application, this can be sent to us using an online file transfer apps such as dropbox or we-transfer. Your portfolio

should ideally include a table of contents with information regarding the duration of each project and in which academic year/level it was completed (predominently from your third year), some design process, design strategy, design resolution and technique strategy and resolution plus a selection of projects worked on during your year in practice.

最少35页的作品集，可以通过dropbox或we-transfer的在线形式提交。目录包含每个设计的跨度、设计时间以及当时的学年（建议主要以大三以上的设计为主）。包含设计过程、设计策略、设计解决方案以及合适的实习作品。

曼彻斯特要求35页以上的作品集，但是许多同学的申请作品并没有这么多页数也顺利被录取了，这是和排版的紧密程度相关的。所以同学们不用担心。如果按照35页以上的排版的话，就需要将很多过程、设计策略分析图放大来展示。

二、澳大利亚

悉尼大学作品集要求

以悉尼大学建筑学作品集要求为例

The portfolio should include examples of coursework submitted for assessment in previous undergraduate courses.

The objective of the portfolio is to demonstrate the level of knowledge and skills attained across all four key streams of architectural study:

– architectural design;

– architectural history and theory;

– architectural technology, including structures, construction, materials and environmental systems;

– design and architectural communication, including manual and digital/software applications for architectural drawing, diagramming, rendering and physical modeling.

Ideally, an example from each stream for each semester of previous study should be included. Where projects were completed as part of a group, applicants should clearly state the role they played in the group and the individual material they contributed to the overall project.

Where professional practice projects are included, the portfolio should clearly and precisely identify what components of the design schemes shown, and what roles the applicant undertook in the project. Reference letters should support this aspect of the portfolio.

作品集应当主要包括四个方面：
（1）建筑设计。
（2）建筑历史与理论。
（3）建筑技术、包括结构、构造、材料和环境系统。
（4）展示手绘和计算机的设计交流技法，例如手绘图纸、建筑图纸、分析图、模型等。

如果是小组设计作业，应当注明申请者在小组合作中的分工，哪些是申请者独立完成的部分。如果有实践作品同样应当标注申请者所负责的具体部分。

新南威尔士大学作品集要求

A portfolio of their design work. A portfolio in digital format is preferred but hard copy portfolios will be accepted. The portfolio should include sample works from various stages of their first degree and text should accompany all drawings/images to explain the projects. Professional work can be included, but the degree of responsibility of the work must be stated.

新南威尔士大学作品集要求很简单，并没有过多的限制：

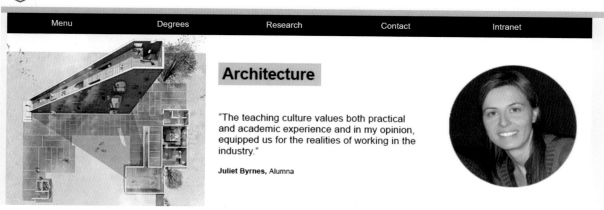

电子版或者打印版的作品集均可（更青睐电子版的），作品集应当包括本科时期不同阶段主要的设计作品。所有的图都需要有相应的文字来对照说明。可以加入实际工程的设计，但是需要说明申请者的分工和主要负责的部分。

墨尔本大学作品集要求

墨尔本大学设计学院作品集要求

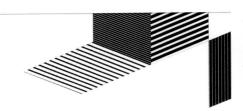

Melbourne School of Design

Design Portfolio Template

This template is for students submitting an application for the Master of Architecture 200pt program. Each page outlines the information that should be included for each project. The information and images of work can be depicted in any style or arrangement, we encourage student creativity.

PORTFOLIO REQUIREMENTS

- The file size of your portfolio should not exceed 10 MB
- The number of pages should not exceed 15
- Recommend that students showcase a minimum of 3 projects. With preferably 4 pages or more per project
- The portfolio must be saved as **one PDF** file and must be in **landscape A3 format**
- Note that portfolios are evaluated digitally and should be easily readable with limited zooming or scrolling

PORTFOLIO EVALUATION

Please refer to the Design Portfolio Matrix below to see how design portfolios are assessed by the Selection Committee.

DESIGN FOLIO MATRIX	
CRITERIA	
1 Evidence of 3-dimensional thinking	
a	Design control over a number of projects of different sizes (small, medium, large), different sites and different programs
b	A variety of spatial representations such as written texts, drawings, physical models and virtual models. (Physical models as photographs)
c	Coherent and readable layout of folio sheets, with clear relationship of text and image (no extraneous graphic effects)
2 Evidence of conceptual thinking	
a	The invention of appropriate compositional procedures and demonstration of use
b	Ability to generate options and select appropriate ones to develop
c	Coherent image/description of each design proposal
3 Evidence of creative thinking	
a	Degree of resolution of pragmatic factors such as site, program, circulation and spatial dimension
b	Degree of resolution of design
c	Expression of enjoyment and delight in the design process

（1）作品集大小不超过 10MB。
（2）不超过 15 页（不是 15 张）。
（3）最少 3 个作品，建议每个作品 4 页左右。
（4）A3 横版 pdf 格式。
（5）作品集应当易于缩放及翻页浏览。
包含不同种类不同规模的设计作品，鼓励不同的表现形式（计算机制图、手绘、模型等）。图文并茂，相互配合。

三、美国

麻省理工学院作品集要求

Portfolio

A digital portfolio is required of all MArch applicants, including those who do not have a previous architecture degree or background. The portfolio file should be exported as PDF for screen viewing. The file should contain no more than 30 pages with a file size not larger than 15MB. Two-page spreads are allowed, but each spread counts as one of the 30 pages.

The portfolio should include evidence of recent creative work, whether personal, academic or professional. Choose what you care about, what you think is representative of your best work, and what is expressive of you. Work done collaboratively should be identified as such and the applicant's role in the project defined. Name, address and program to which you are applying should also be included. We expect the portfolio to be the applicant's own work. Applicants whose programs require portfolios will upload a 30-page Maximum), 15MB (Maximum) PDF file to the online application system. The dimensions should be exported for screen viewing. Two page "spreads" are counted as one page.

作品集要求：

（1）电子版PDF格式作品集不超过30页，大小不超过15MB。

（2）用最近的作品来展示申请者的创造力，个人的、学术的或者实际工程作品均可。

（3）展示最能代表作者设计思想的作品。

（4）合作作品需要标注负责的部分。

（5）姓名，地址以及申请专业需要注明。

哈佛大学作品集要求

Applicants to degree programs other than the Master in Urban Planning program must submit a portfolio that includes their most important and representative visual arts, design, research, and/or professional work. A portfolio for the Master in Urban Planning degree program is optional. Master in Design Studies and Doctor of Design portfolios should consist of scholarly, academic and/or professional work and may or may not include visual material, at the discretion of the applicant, and as related to the proposed research area.The design portfolio for Master in Design Engineering applicants should provide three to five examples of the candidate's work, design and/or research, which are most relevant to the MDE program. Examples may be professional or academic and may encompass designed, researched, and/or actualized work. Projects

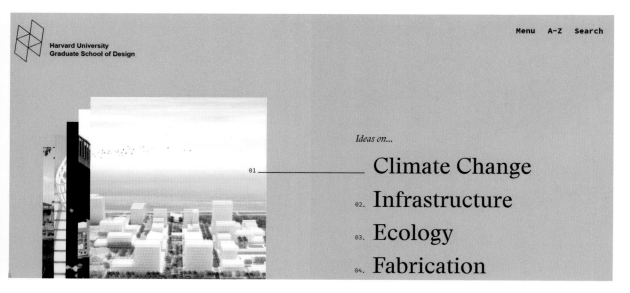

may include, but are not limited to, buildings, mechanical systems, electronic systems, organizational systems, and recommendations regarding processes. Demonstration of technical literacy and innovation is desirable. Projects should be submitted in PDF format and include a description of the project as well as supporting documentation, which may include images or a video (videos must be uploaded separately from the PDF). Applicants may include a list of additional relevant projects.Please see admission instructions for more details on portfolio submittal. Special students may also be required to submit portfolios. All portfolios must be submitted online; hardcopy portfolios will not be accepted.

作品集要求：

除了城市规划专业，其他哈佛大学学院的课程都需要提交作品集。城市规划专业的作品集是可选项而非必选项。设计专业的硕士和博士的作品集需要包括学术或实际工程作品。作品内容应当与申请方向相关。设计工程硕士方向的申请者应当提供3到5个设计作品，可以包括建筑、机械系统、电子系统等相关的领域。作品集如果包括视频，视频文件应当单独上传。不接受打印的作品集。

宾夕法尼亚大学作品集要求
Architecture and Landscape Architecture Portfolios

Every applicant to the Master of Architecture, Master of Science in Design, Master of Environmental Building Design, Master of Landscape Architecture, Master of Science in Architecture, or PhD in Architecture program is required to submit a digital portfolio. Paper portfolios will NOT be accepted, and if submitted, will not be returned. The portfolio is a synopsis of one's creative work. As a visual essay, it tells a story of a person's interests, skills, and development over time. It should include projects that best express one's visual, spatial, and constructional abilities. These projects might include drawings, paintings, sculpture, or photography; graphic, industrial, or interior design; architectural, landscape, or urban design. The faculty who evaluate the portfolios look less for competence in architectural or landscape architectural design and more for a coherent demonstration of visual and spatial abilities expressed through a basic understanding of material and construction. Applicants to the MEBD and MSD should include at least five fully developed projects done solely by the person submitting the portfolio; other group work can be added.

The digital portfolio should be formatted as one PDF document no larger than 10 MB, with no more than 20 pages (Maximum page size 10

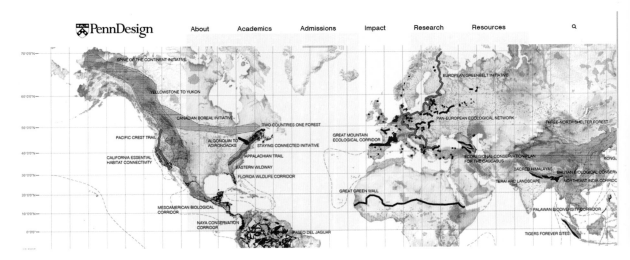

x 12") or 10 pages (Maximum 10 x 24") if you use spreads. Cover pages or table of contents do not count towards total pages. We suggest that you keep the format consistent. For instance, if you are using spreads, use all spreads. If you are using pages, don't mix in spreads. Once your application is submitted, you will not be able to make changes to your portfolio or upload a new one.

City Planning and Urban Design Portfolios

Applicants to the Certificate in Urban Design program should submit portfolios containing reproductions of their work in the visual arts and design. Applicants to the Master of City Planning program with a concentration in Urban Design are strongly encouraged to submit a portfolio. Applicants with professional experience who wish to include examples of their professional work may do so, but are advised to limit these to projects in which they had principal design roles. If examples of collaborative projects are submitted, the applicant's contribution should be specifically described. The digital portfolio should be formatted as one PDF document no larger than 10 MB, with no more than 20 pages (Maximum page size 10 x 12") or 10 pages (Maximum 10 x 24") if you use spreads. Cover pages or table of contents do not count towards total pages. We suggest that you keep the format consistent. For instance, if you are using spreads, use all spreads. If you are using pages, don't mix in spreads. Once your application is submitted, you will not be able to make changes to your portfolio or upload a new one.

Fine Arts Portfolios

A portfolio is required of all applicants to Fine Arts, Time-Based and Interactive Media, and Emerging Design and Research programs. The portfolio should indicate your major interest, represent your best work, and demonstrate your abilities. At least half of the portfolio should consist of work completed in the last two years. Applicants may choose to provide still images and/or video. All work in the portfolio, including images and concepts, must be original material created by you, and should be identified as academic, professional, or personal. If professional or team projects are included in the portfolio, you must clearly identify your specific role and responsibility in the production of the project. Labels and writing should clearly explain the work.

建筑与景观建筑学作品集

作品集应当体现申请者的兴趣，技能以及能力的变化发展。

作品集做成 PDF 格式，大小不超过 10MB。用单页排版页面大小不超过 10 英寸 ×12 英寸，20 页以内。用连页排版，页面大小不超过 10 英寸 ×24 英寸。封面与目录不算在页数之内。不建议单页排版和连页排版混用。

城市规划与城市设计作品集

申请者的实际工作经验可以加入作品集中，但是需要注明申请者所做的共享范围。强烈建议城市规划专业的申请者也提交作品集（特别是如果学习方向偏向于城市设计）。作品集做成 PDF 格式，大小不超过 10MB。用单页排版页面大小不超过 10 英寸 ×12 英寸，20 页以内。用连页排版，页面大小不超过 10 英寸 ×24 英寸。封面与目录不算在页数之内。不建议单页排版和连页排版混用。

艺术作品集

作品集应当体现申请者的兴趣，技能以及能力。至少有一半的作品内容应当是近两年完成的。原创的图片和视频都可以包括在内，如果有实际工作作品或是合作作品，个人的角色和负责内容需要清楚陈述。

四、北欧

瑞典皇家理工学院作品集要求

1. Format: PDF

2. Size: Maximum size of 10 MB, no more

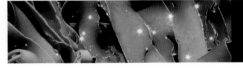

than 7 pages+1 page cover sheet, single-sided, not spreaded.

3. Minimum 2-3 projects, preferably no group work.

4. Anonymous portfolio: Do not include your name anywhere in the portfolio. Include your 7-digit application number on the first page of the portfolio instead.

5. File name: The file name should be the same as the application number. Example: 7538761.pdf

Do not mix up the 7-digit application number with the programme code (E0121).

6. Content: The portfolio should only present the applicant's educational projects, no projects from work experience/internship are accepted. The projects included should be correctly titled:

7. State during which year of the studies the project was done

8. Indicate if the work has been produced by the applicant alone or in collaboration. Please make sure to strictly follow the instructions above otherwise your application will be marked as ineligible.

建筑学院作品集要求

（1）PDF 格式。

（2）10MB 以内，7页加1页封面，单页开。

（3）至少2到3个设计，最好都是独立作品。

（4）作品集匿名，在封面标注7位数的申请号。

（5）作品集以申请号命名。

（6）作品集应当包括申请者的主要学术作品，不接受实际工作的设计作品。

（7）标注作品的完成年份。

（8）标注作品是独立完成的还是合作作品。

阿尔托大学作品集要求

A portfolio of the applicant's own practice work and any other personal work is a compulsory and important appendix to the application. It must illustrate the applicant's architectural design skills and artistic nature in a broad-based manner. The main focus of the evaluation is placed on design sketches and other image material.

阿尔托大学作品集要求：包含申请者的独立作品。必须展示申请者的设计技巧、美学修养和广泛的能力。设计草图和设计图将被作为作品集评判的主要依据。

意大利米兰理工大学作品集要求

The portfolio must be uploaded on your online profile in the section "Master of Science programme(s) selection" and cannot be bigger

than 15 MB. Please do not send hard copies by post since they will not be considered. The portfolio is a collection of selected samples of work from your previous educational programme or work experiences. It should reflect the scope and variety of your training and experience with specific emphasis on your knowledge of and skills in the degree track for which you are applying. You should submit works that represent your interests and personal views as well as your design and technological abilities. Neatness and clarity of presentation are extremely important as they reflect both your attitude towards your work as well as your ability to communicate your work in a comprehensive and deliberate manner.

米兰理工大学作品集要求：在申请系统里上传电子版，不大于15MB。作品集是包括学术和工作作品的选集，应当体现申请者知识与技能的广度和深度。作品需体现申请者感兴趣的学习领域。作品集的整洁性是非常重要的，整洁性反映了申请者的态度和通过设计交流的能力。

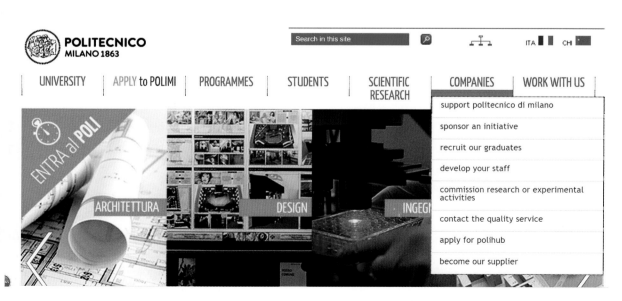

荷兰代尔夫特理工大学作品集要求

Number and nature of projects:

● Include a total Maximum of 5 projects.

● Include no more than 2 professional projects.

● Include at least one individual project.

● Include your final project (if you have not yet completed it, include your most recent project).

● At least one of the projects included should clearly demonstrate your capabilities with respects to structural and technical project viability. We highly recommend that these structural and technical aspects are exhibited in either the final thesis project, or alternatively, your most recent or most advanced level, project.

● Include a summary of your motivation letter of no more than 400 words (make sure all elements of the essay are covered in this summary).

You should present your best (and/or most thorough) projects and describe them in detail, graphically demonstrating your ability to initiate and follow through the design process.

Per project include:

● concept sketches, showing preliminary visual (and verbal when relevant) development of ideas, approaches and methods and showing how you organise your ideas.

● finished drawings of original design work, documenting in a clear and precise manner both the intention and the resolution of each project.

● a written explanation (in English) of the design concept and solution Here you should include the inspiration behind the project as well as the goals and objectives of the design. Be clear and concise. Longer texts should be worked out in the form of your essay.

Please note that for all drawings we require you to include the name of the author.

Size

● The size of your portfolio should not exceed 20 MB.

● The size of the pages should not exceed A4.

● The number of pages should not exceed 30 A4 size pages.

● It is not allowed to show more than one A4 size page per screen

● It is not allowed to use the booklet format because that is two pages per screen

● The format should be pdf.

● Be aware that a landscape layout suits our requirements best.

作品集基本属性：

（1）最多5个设计作品。

（2）不超过2个实际工程设计。

（3）至少有1个独立完成的作品。

（4）包括你的毕业设计（或者最近的设计）。

（5）至少有1个设计体现结构和技术的知识（比如毕业设计或者最新的设计）。

（6）包括1个400字以内的动机信。

申请者应当提供最好的设计，通过详细的图示语言展示设计的思路和过程。

每个设计的要求：

（1）草图，方案发展的过程，对想法的归纳组织。

（2）完善详尽的最终图纸，用清晰有逻辑的方式呈现。

（3）相应的解说文字，包括理念阐述和灵感来源。

格式：

（1）不超过20MB。

（2）版面不大于A4。

（3）不超过30页。

（4）A4单页排版。

（5）不要用书籍的竖版双页对开排版。

（6）PDF格式。

（7）横版排版最符合我们的要求。

五、中国

香港大学作品集要求

Design Portfolio (A4 landscape orientation in PDF format; 20 pages; 30 MB file size limit): Including academic design work and professional work;

香港大学的作品集要求比较简洁：A4横版PDF格式，不超过20页，大小不超过30MB，包括学术作品和实际工作作品。

由此我们可以看出，制作作品集之前一定要做好功课，才能有好的设计，好的表达。还要按照学校的指定要求来制作，才能保证最好的申请结果。A4横版是大多数学校的最爱。作品集的大小也是同学们需要格外注意的，大部分同学的作品集初稿完成之后都是2300MB甚至12GB大小。因此需要在尽量保持图像质量的同时压缩作品集（压缩到十分之一甚至百分之一的大小，像素损失是难免的，只要在可以接受的范围内即可）。最专业的PDF制作、修改和压缩软件是Adobe公司出品的Adobe Acrobat。同学们可以根据自己的需求来进行作品集的制作和压缩。

后记

本书的初衷是为了给热爱设计，有着坚定留学梦想的设计行业学子们起到一个抛砖引玉的作用。让大家对留学作品集有一个初步的了解认知，熟悉作品集的内容，内在规律，申请的规则。避开常见的误区，为申请者准备作品集素材，确定方向节约宝贵的时间。

本书包括的大量内容都是基于笔者近年来的作品集辅导和申请辅导经验，实际的辅导经验和与学校直接的沟通交流使本书具备最新的时效性和极好的实用性。所有的设计图例都选自己被世界一流名校录取学子的作品集。这些作品的作者大都已经在境外名校继续攀登人生高峰，走在实现自己梦想的道路上了，也有很多同学已经顺利毕业，在著名设计公司里施展自己的抱负。我非常欣慰地看到他们的成长和进步，他们现在的设计已经更上一层楼，在这里感谢他们的信任与大力支持。本书的完成离不开巅峰建筑学社另外两位灵魂人物：金梦潇女士和郑权一先生的鼎力支持。没有他们的支持，本书的完成是难以想象的。他们对教育事业的热爱和倾心付出深深感染和激励着我，十年树木，百年树人，我们愿为广大有梦想，有热情，为自己人生努力奋斗的同学们助力。

在这里，我想用百年前梁启超先生《少年中国说》里的名句来做此书的结束语："红日初升，其道大光；河出伏流，一泻汪洋。潜龙腾渊，鳞爪飞扬；乳虎啸谷，百兽震惶；鹰隼试翼，风尘吸张。奇花初胎，矞矞皇皇；干将发硎，有作其芒。"当今之中国，无数学子怀抱梦想远渡重洋，等待你们的必将是辉煌的明天！今日之中国少年，必将成为他日世界之栋梁！

向畅颖
于丹麦奥尔堡